GENES UNLEASHED:
How EPIGENETICS Controls our Genes

Joyce Taylor Papadimitriou
Joy Burchell

CONTENTS

Acknowledgements

We would like to thank all the scientists and clinicians who have worked in the field of Epigenetics for publishing their work, thereby making the research for this book possible. We would particularly like to thank Dr Richard Beatson for reviewing drafts of the manuscript and providing excellent and useful comments.

Writing this book took longer than we had initially anticipated, and we wish to thank all members of our families for allowing us the time to focus on the book and for the endless discussions on the subject.

About the Authors

Joyce Taylor-Papadimitriou (MA Cantab; PhD Toronto) is a Visiting Professor at King's College London (KCL). She was born in Burnley, Lancashire, and has authored over 300 scientific publications, the most recent being in the field of Epigenetics. After meeting and marrying Spyros Papadimitriou, she moved to Greece, where she lived and worked for 10 years. Upon returning to England, she was recruited to the Imperial Cancer Research Fund (ICRF) and began a collaboration with Joy Burchell, which has lasted 40 years. As an elected Fellow of the Academy of Medical Sciences (FMedSci), she is among the most influential biomedical scientists in the UK and is recognised globally.

Professor Joy Burchell is Professor Emeritus at King's College London. She obtained a PhD in Cell Biology while working at the Imperial Cancer Research Fund under the supervision of Professor Taylor-Papadimitriou. Within Joyce's lab, she became a Staff Scientist and, in 2004, obtained an independent position as Senior Scientist at Cancer Research UK. In 2007, she moved to King's College, and in 2013 she was awarded a Professorship. She is an internationally recognised expert in the involvement of sugars in breast cancer and the cancer microenvironment. She has published over 150 highly cited research papers in prestigious journals, including articles on Epigenetics.

Introduction

Have you ever wondered how the cells and organs of your body can perform such different functions, and - as long as you are healthy – get it right? Since we have learned more about our genes being in DNA, and that all our cells carry the same DNA sequence but do different things, this fact has become an even more awe-inspiring truth. We do not make teeth in our liver or nails in our gut, so how does this system work? Trying to answer this question constitutes the study of Epigenetics. This book is designed for the curious, who want to learn more about discoveries made by scientists trying to answer the question of how the expression of different genes (whether genes are switched on or off) in different tissues is controlled.

Genetics of course, as the name suggests, relates to the study of genes, which we now know are made up of DNA, and our ancestry can be studied by following the transmission of our genes down the different generations. The word Epigenetics was coined long before we had the sequence of our DNA and was suggested by a visionary (Conrad Waddington) using the Epi prefix from the Greek which means above and beyond (genetics). He knew that the genes were crucial and predicted what we now know to be true - that the control of their expression is basic to the development of the different tissues, both in embryonic development and in the adult. The epigenome is the term which refers to modifications to the genome (the complete set of genes present in the cell) that do not affect the DNA sequence and defines the profile of genes expressed or not expressed. This is clearly different not only for each tissue but for the stage of development of that tissue. We hope that the ensuing chapters will draw you in to one of the most exciting fields of scientific enquiry, that is now beginning to provide answers to some of the most fascinating questions in biology while at the same time providing a magical aura for the human condition.

Clearly, in this book, there has been selection of content, emphasising simplicity while retaining accuracy, and some readers may be frustrated by the limited nature of the information, while other readers coming fresh to molecular biology may be dissatisfied for other reasons. To comprehensively cover the whole new world of Epigenetics would require a whole library devoted to the subject. We have tried to optimise the size of the content here to whet the appetite of anyone who is curious, and we anticipate that the glossaries for each Chapter will be useful for reference to unfamiliar words.

The first two chapters aim to provide enough basic information to be able to digest the mechanisms which are unfolding regarding how regulation of gene expression is achieved. We begin in Chapter 1 to explore the structure of the DNA in our chromosomes (which are found in the nucleus), which makes up 99% of our DNA and where studies led to the identification of the "double helix" as the basic structure. It is important to recognise that, in the context of Epigenetics, the basic unit of this genome is not DNA alone, but chromatin, which is made up of DNA wrapped around proteins known as histones. Moreover, for the epigenetic changes occurring in normal development, where, in each tissue, the cells "differentiate" so that they can perform the function required by the tissue, it is the addition of small chemical molecules to DNA itself, or to the histone proteins, which affect gene expression.

The word differentiation describes the normal epigenetic pathway of cells in a tissue as they change their epigenome to be able to carry out the functions of that tissue. Keeping the balance between producing more cells (proliferation) and replacing dying differentiated cells, which cannot divide and only survive for a short time, is crucial for the dynamic stability of the tissue. The description of chromatin is presented in Chapter 1, and just how the chemical modifications of DNA and histones can affect gene expression are discussed in Chapter 2.

While 99% of our DNA is found in our chromosomes, the other 1% is found in structures called mitochondria, which are also found in our cells but outside of the nucleus. While this may seem to be a small percentage, these structures and their DNA are vital to our well-being as they produce the energy we need from the food we eat. In Chapter 3 we describe these mitochondria, which were originally not ours, but developed from bacteria which invaded our cells, around 1.5 billion years ago, and stayed as house guests. These bacteria brought with them an amazing system that uses the oxygen in our atmosphere to produce the energy needed for all our activities. Unlike chromosomal DNA which we inherit form both parents, surprisingly, we only inherit our mitochondrial DNA from our mother. Nevertheless, over the years both forms of DNA have learned to interact and to epigenetically control the function of each other. Moreover, both have proved to be very useful in forensic science and for identifying unknown and ancient remains.

It is becoming very clear that changes in the epigenome are associated with several diseases. As an example, we have focused on cancer in Chapter 4, and to a lesser degree on mitochondrial diseases in Chapter 3. Cancer can develop in so many different tissues and is a proliferative disease showing an imbalance of proliferation versus differentiation. Many therapies used in cancer try to drive the cancer cells to differentiate rather than proliferate. Changes in the epigenome pathway play an important role in the development of cancer and are potentially reversible, therefore new therapies based on these changes have been developed and are now being applied in the clinic for cancer therapy. This is discussed in Chapter 4, as well as the use of novel methods of diagnosis based on epigenetic change.

Since epigenetic change is part of the normal life of cells, and these changes are induced by signals generated within the organism, the question arises as to whether signals from the environment can also induce epigenetic change. This is very relevant to the overall view of evolution and whether the Darwinian principle of random mutations of genes which happen to give an advantage for survival are the only factors affecting selection of the fittest. Certainly, Homo sapiens has been able to enter and survive in nearly all of earth's different environments, and it is a valid question to ask if epigenetic change can play a role in the survival of the species. This question separates into two parts, the first relates to whether the epigenome of an individual is affected by an environmental stimulus, the second- whether the change in this individual can be passed to his offspring. The second part of the question asks whether an epigenetic change is "transgenerational", which it would have to be to affect the survival of the species.

One of the most important stimuli from the environment is contained in what we eat. There have been studies which have observed how involuntary changes in diet, caused by natural events or war, can affect the individual and their offspring. These usually relate to food restriction (famine) or the availability of an abundance of food at different stages of the life cycle of Homo sapiens. This is discussed in Chapter 5 where the studies of the Dutch Hunger Winter, examining the health of individuals exposed to famine in the womb are described. There is less data from this study relevant to part 2 of the question asking whether epigenetic changes can be transgenerational. However, a Swedish study following the health of individuals exposed to famine or to abundance of food at the age just before puberty, has yielded tantalising data suggestive of transgenerational inheritance and this is also discussed in Chapter 5.

The above studies are termed observational studies as we the observers have no direct effect on the changes seen. It is difficult to make controlled experiments with humans on the effect of diet on the epigenome but studies in mice - our fellow mammals- have pointed the way to considering important features of diet. One such feature is the realisation that it is important to feed, not just ourselves, but also the bacteria, i.e. the microbiota, that share our body, (predominantly in the gut) in order to maintain the symbiotic relationship we have with them. Unlike the mitochondria, which developed from invading bacteria and became an intrinsic part of our cells, the bacteria in the microbiota are living in symbiosis with us. We provide them with food, and they contribute to our health. They use the amino acids which we produce from the proteins we eat and digest, and the fibre in our diet, which we can't digest but they can. For their part, they also make factors that we need for optimal health, especially in the gut. We consider this in Chapters 6. In Chapter 7 we discuss the importance of the spectrum of the bacteria making up the microbiota in health and in some diseases. They, the microbiota, are extremely important for a healthy immune system and this dependence is evident in inflammatory bowel disease and in the forms of cancer therapy which aim to reverse the negative effect which cancer cells have on the immune response. Indeed, the profile of bacteria carried by an individual can predict their response to some forms of immunotherapy, and transferring the microbiota from one individual to another is now a possibility. All of this we discuss in Chapters 6 and 7.

While the study of the microbiota and their effects on the epigenome is a very active field of research, it will be important for us to be aware that characterisation of what is a "beneficial" bacterium and what is a "non-beneficial" bacterium is still a research activity. Nevertheless, it is clear that a broad spectrum of bacteria in the gut is complementary to good health, and the addition of food containing probiotics may be advantageous. We should however always be wary of competing advertisements for yoghurts (probiotics) suggesting that these foods are the answer to a healthy gut microbiome. Instead, we should consider that the best way to maintain a healthy gut microbiome is to have a varied diet with plenty of fruit and vegetables, and fibre - to feed both ourselves and our microbiome. With the research community continuing to study and define beneficial bacteria, the use of non-food probiotics have the potential to be very important, not just for health but for the treatment of some diseases.

We hope this book will give you an insight into the fascinating world of Epigenetics and perhaps awaken your curiosity to learn more. We hope it might go some way to explain why individuals with exactly the same genetic code, that is identical twins, do have differences in their response to their environment and in fact can develop different diseases. Our whole existence is dependent on our epigenome, and we would like to think that we have opened up vistas which display not only a glimpse of what is already known, but also how much there is still waiting to be learned about Epigenetics.

We finish this Introduction with some words from

Emily Dickinson*:*

Exultation is in the going
Of an inland soul to sea
Past the houses - past the mountains
Into deep Eternity—
Bred as we are among them mountains –
Can the sailor understand
The divine intoxication of the first league out of land?

While it is too much to expect exultation and definitely not intoxication, but the analogy of moving from the limited land to the expansive sea can, if we like, signify the huge possibilities of moving forward with the prefix - above and beyond - to take part in:

"The New World of Epigenetics."

Chapter 1
Genes and DNA: The Mystery of Coding

"Look deep into nature, and then you will understand everything better"

Albert Einstein

We have all heard of DNA, indeed it has entered into everyday language with sayings such as "it's in your DNA" but what actually is DNA and how does it code for our genes? Indeed, what are our genes? This used to be easily answered as stretches of DNA that code for proteins but within recent years it has become clear that there are also many genes which do not code for proteins (termed non-protein coding genes). These genes in fact are expressed and code for an RNA which has a function of its own. In this chapter we will focus on considering the molecular events which lead from a protein-coding gene being expressed (transcription) to the production of a protein (translation), leaving the discussion of the non-protein coding genes for the next chapter.

Before we go on to show how epigenetics can control gene expression it is necessary to understand DNA. Our bodies are made up of cells (roughly 30 trillion) and in "eukaryotes" like ourselves (see glossary), DNA is mostly found within the nucleus of the cell, which is like a porous bag placed within material known as the cytoplasm. Although the bulk of DNA is found within the nucleus, known as nuclear DNA (nDNA), and most of this chapter will be devoted to the structure and function of this DNA, structures called mitochondria, found in the cytoplasm, also contain DNA (mtDNA) and we will briefly describe these later in this Chapter. A more extensive discussion of the function of the mitochondria will be presented in Chapter 3, after we have talked about some of the mechanisms underlying epigenetic change in Chapter 2. Before we describe the structure of DNA and how it codes for the proteins, (which are the molecules that do a lot of the work in a cell) a knowledge of some of the history of the earlier studies in biology which led up to the identification of DNA as the carrier of our genes are of great interest. These studies illustrate how the culture of society affected the scientists working at the time and how the hypotheses they developed were regarded by Society.

The Road to Discovery of Genetic material

Charles Darwin and the Origin of Species

It is now difficult to comprehend the depth of the antagonism which arose in the educated classes of society to the ideas that were expressed in the book "The Origin of Species" by Charles Darwin, which was published in1859. The book put forward the theory that all species of organisms arise and develop through the natural selection of small, inherited variations that increase the individual's ability to compete, survive, and reproduce. Of course, it was not known what the molecular changes were which led to the changes in behavioural patterns of the different species, but it is now clear that these changes are occurring in the genes found in our DNA.

Darwin's ideas disturbed the creationist view that all species were created by God and were immutable. Darwin himself knew of the controversial nature of the ideas, since he was trained at Cambridge University and was aware of the power which the Anglican Church exerted over all walks of life. Until the early part of the 19th century there were only two universities in England, Oxford and Cambridge, and Darwin's friends and colleagues were elite gentlemen biologists with appointments in the Anglican Church. Indeed, scientific research at this time was carried out by gentlemen with positions in the Anglican Church or friars and monks in religious orders. It was not until the University of Durham was established in 1832 and University College London (UCL) and King's College London (KCL) were established in 1836, that scientists could have a scientific career with a salary paid by the Government. The establishment of these universities was followed by the formation of many other Universities in the UK liberating science from the disapproval of religious bodies.

Looking back at Darwin's recent ancestry, it is surprising that he was not ready to go public as a free thinker and held back for many years from publishing the data he had collected, documenting the different life forms (i.e. species), when he went around the world for five years on the Government sponsored ship The Beagle. Charles's grandfather from his father's side (Erasmus Darwin) was a famous doctor who was a free thinker and a poet who said, "*What need of Christianity when we can sup the milk of science*" and wrote:

"*Nurs'd by warm sun-beams in primeval caves*

Organic life began beneath the waves...

Hence without parent by spontaneous birth

Rise the first specks of animated earth".

Darwin's grandfather through his mother's side was Josiah Wedgewood who founded the Wedgwood Pottery Dynasty and who adopted the Unitarian version of Christianity which was in opposition to the teachings of the Anglican Church.

Erasmus Darwin and Josiah Wedgwood were great friends, and both heartily approved of, and were involved in, the Industrial Revolution. The Darwin and Wedgwood families had intermarried down the years and in fact Charles's wife Emma was a Wedgewood. So why were the liberal views and lifestyles evident in the age of "Enlightenment", which were operating in the lives of the grandfathers, not progressing further – in Charles Darwin's time? The reason lies in the effect that the French Revolution had on society in Britain, where a period of repression set in, and radical dissenters and free thinkers were attacked for disturbing the status quo. Charles and his "Origin of Species" were faced with a reactionary society, antagonistic to liberal ideas, Moreover, he had within himself an internal dilemma associated with the conflict of the religious views with his scientific data. This may have contributed to the ill health he experienced all his life and restrained him from beginning the writing of "The Origin of Species" for many years. He was however spurred into beginning the writing when he received a letter from Alfred Wallace.

Wallace was a socialist, with none of the privileges which had supported Darwin during his life, and he had also been studying the different species of life throughout the world, supporting himself financially by collecting specimens and selling them to Institutes and even to Darwin. The contents of the letter from Wallace made it clear that he had arrived at the same concept about the origin of species as Darwin. Colleagues of Darwin then arranged that both scientists would present a short paper to the members of the Linnean Society describing their findings, with Darwin being the first speaker. However, these papers were read to a limited audience and were later dominated by the extensive publication of the "Origin of Species" by Darwin.

Getting closer to the source of the genetic material

At roughly the same time as Darwin published his book, in a monastery in Brno, now in the Czech Republic, an Augustinian monk called Gregor Mendel, derived the law of inheritance by breeding peas in a small garden of the monastery[1]. At the time there was commercial interest in understanding how characteristics were inherited from one generation to the next because of the breeding of sheep for wool production, and the Friary interacted with local universities and the famous University in Vienna to study this problem. This sort of interaction goes on today where companies can fund universities or professors to collaborate, working on a scientific problem.

So, Mendel, who was a good mathematician, was selected to receive some training in Vienna and he decided to work with peas as a model system to study this problem of inheritance. The monastery gave him a very small patch of land for his experiments. He selected certain traits such as colour and texture which he could observe and follow. By carefully crossing the peas and keeping good records, he showed that the colour of the peas could be inherited from generation to generation, some in a dominant way, some in a recessive way. When Mendel crossed a green pea with a yellow pea the offspring were always yellow, thus the yellow colour is dominant and green recessive. However, in the next generation, obtained by crossing the yellow peas produced in the first generation, the green peas reappeared - at a ratio of 1:3, one green pea, to three yellow peas (see Figure 1.1).

Mendel published this work in 1866 showing that "invisible factors" could determine characteristics of organisms. These invisible factors are what we now know as genes. What Mendel showed is that characteristics could be inherited and that some characteristics are dominant, that is only one copy of the gene is required for the characteristic to be expressed, and some are recessive, where two copies are required for expression. Mendel's results with the peas can explain how children from human parents where both carry a disease-causing *recessive* gene (DCG), can have children who do not have the disease (carrying one copy DCG) and children who do (with 2 copies of DCG). They also explain why too much intermarriage between cousins may not be good idea, if both parents carry the bad gene. Mendel's crucial contribution to our understanding of genetics was not really recognised during his lifetime and he only posthumously gained the title as the founder of genetics in the 20th century.

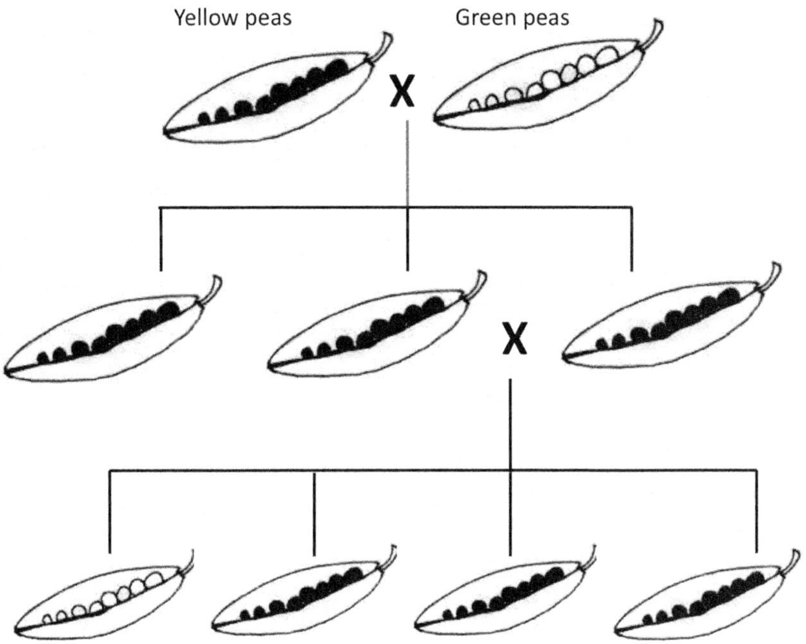

Figure 1.1: Mendel's pea experiment. *The gene coding for yellow peas is dominant so when a yellow pea is crossed with a green pea all the offspring are yellow. However, if these 2nd generation yellow peas are crossed then a 3:1 ratio of yellow (3) to green (1) will be obtained. The gene for the green colour is recessive.*

Identification of DNA as the Genetic Material

Fredrich Miescher was a Swiss doctor working in the laboratory of Hoppe-Seyler in Germany. He was tasked with isolating white blood cells and found that the best source of these was from pus seeping from infected wounds! He therefore collected bandages covering such wounds and isolated the cells. In 1871 he published a paper describing the isolation of "nuclein" from the nucleus of these cells which was composed of DNA and protein. Although Miescher showed that both protein and DNA were present in the nucleus of the cell, it was unclear which of these was responsible for coding our genes. We take a great leap forward now to the 1940s. In 1944 a Canadian born scientist named Oswald Avery working at the Rockefeller Institute in New York, demonstrated that hereditary changes could be induced in bacteria by transferring DNA into the bugs[2]. This showed that DNA is the bearer of the genetic specificity. Although now thought of as one of the most important papers of the 20th century, Avery was never awarded the Nobel Prize although he was nominated a number of times.

The structure of nuclear DNA

We now move to the most well-known discovery in the field of genetics, that of the structure of nuclear DNA proposed by Watson and Crick who, at the time, were based at the Cavendish laboratory of Cambridge University in the UK. By the mid 20th century science was flourishing and Watson and Crick were not impeded in their research by religion, as was Darwin.

It is interesting to note that the two scientists, who met in 1951, began their collaboration when they were at very different stages of their careers and how chance, plus connections, led to their meeting. Francis Crick was a physicist with a degree in physics from University College London but his studies working for a PhD degree were interrupted by the second world war when he was recruited to do research for the Admiralty in the design of magnetic and acoustic mines. After the war, at the age of 31, he decided that he wanted to work in biology and with this aim in mind, he obtained a studentship to work at the Cavendish laboratory in Cambridge, where he then began the research which led to his being awarded a PhD in 1954 (at the age of 38) - one year after publishing the famous paper with Watson describing the structure of nuclear DNA.

Watson on the other hand seems to have been a boy wonder, obtaining a first degree in Zoology at age 19 (in 1947) and a PhD at the University of Indiana Bloomington aged 22. There he acquired an interest in genetics, influenced by, the well-known microbiologist, Salvador Luria. It was later - in 1951 - when he was a post doc at the University of Copenhagen, that James attended a conference where, in a chance encounter with Maurice Wilkins (Kings College London), he learned of the work going on in the UK using X rays to study the structure of DNA. He clearly wanted to take part in this research and his mentor from the US, Luria, arranged for him to work at the Cavendish Laboratory in Cambridge. Here Watson the biologist met Crick with his background in physics and they began their collaboration leading to defining the structure of DNA.

In 1953 Watson and Crick published in the prestigious scientific journal Nature[3]. The paper was just one page long and described the structure of DNA as being a double helix. They concluded that DNA was made up of two strands of nucleotides that were held together and twisted into a double helix. However, this was not the first description of Watson and Crick's work as in March 1953, Francis Crick wrote a letter to his 12-year-old son Michael, who was at boarding school, describing their theory on the structure of DNA[4]. In this letter he describes how, "the exciting thing is that while there are 4 different bases, we find that we can only put certain pairs of them together". Francis Crick finishes his letter, "Read this carefully so that you understand it. When you come home, we will show you the model, lots of love, Daddy". It was a lot to take in for a 12-year-old! This letter was sold by auction in New York in 2013 for a record breaking $5.3 million dollars, the money going to benefit scientific research.

If DNA "codes" for our protein-coding genes, it must be made up of a code. DNA, which stands for deoxyribose nucleic acid, is made up of four bases, or nucleotides, known as A, T, C and G. The various combinations of the four bases form a code, but we will come onto that later. So, what was so important in the nucleotides only being able to fit together in pairs? Watson and Crick showed that of the four nucleotides T could only link to A, and C could only link to G. This means that when a cell divides and the double helix unzips into two strands, the new nucleotides required to form new double helices can only be added to their partner nucleotide. This ensures that the cell makes two perfect copies of the original double helix, and one of these can be passed to each of the two daughter cells upon division. Figure 1.2 shows how this can happen.

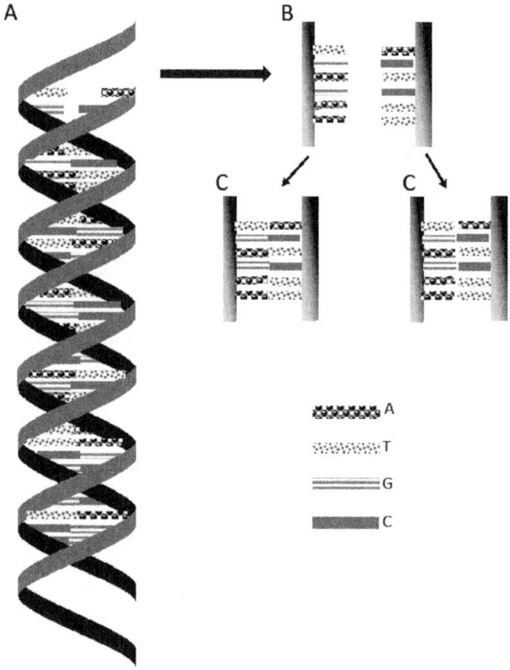

Figure 1.2: DNA forms a double helix and can make a perfect copy of itself. *The backbones of the DNA are twisted around each other to make a double helix, A: When the cell is about to divide the helix can unwind, B: As the base C can only go with base G, and base A only with base T, adding these bases to the unwound strand results in a perfect copy of the parental DNA as shown in C.*

In 1962 James Watson, Francis Crick and Maurice Wilkins were awarded the Nobel Prize in Physiology and Medicine for this work. Maurice Wilkins was an X-ray crystallographer, based at King's College London. Rosalind Franklin was also an X-ray crystallographer who was recruited to King's by the biophysicist Sir John Randall. X-ray crystallography is a technique for determining the structure of a crystal. The principle is that crystalline structures cause a beam of X-rays to diffract into specific directions and by looking at X-ray films of the diffracted X-rays, the structure can be determined. What is absolutely vital to this technique is the preparation of the crystals and Rosalind Franklin was very good at this, she was an excellent chemist. Whatson and Crick were not chemists. Rosalind Franklin with her PhD student, Raymond Gosling, took an X-ray photograph of the crystalised structure of DNA that had been isolated from ram's sperm. The controversy that surrounds this discovery is that it is said that Maurice Wilkin's took the photograph, known as photograph 51, out of Rosalind Franklin's papers without her knowledge and showed it to Watson and Crick. Moreover, a recent comment published in Nature[5] suggests that this was not a "eureka" moment for Watson and Crick. which enabled them to propose the structure of DNA but an important piece of evidence. Moreover, evidence uncovered and reported in the above comment showed that Rosalind "did not fail to grasp the structure of DNA" but was equal in solving it. Data found in a Medical Research Council (MRC) report on the activities of the King's MRC unit which included a report on Rosalind Franklin's work, was written as part of the MRC official visit and was unofficially shared with Watson and Crick by Max Perutz who was Crick's PhD supervisor. These data and the vital photograph allowed Watson and Crick to spend 6 weeks of trial-and-error fiddling with cardboard models to define the structure of DNA. From the Nature Comment, it is clear that Rosalind Franklin was an equal contributor to the discovery of the double helix structure.

The story of Rosalind Franklin was developed into an excellent play in the West End called Photograph 51 starring Nicole Kidman as Rosalind. The play really brings to life the contribution of Rosalind Franklin to the discovery of the structure of DNA and proposes the question, were women side-lined in science? However, Watson, Crick and Wilkins were not awarded the Nobel prize until 1962 and sadly by that time Rosalind Franklin had died of ovarian cancer at the young age of 37. Nobel Prizes cannot be awarded posthumously. Nevertheless, Rosalind will be on an exploratory journey to Mars in the near future with the rocket, which is to be named after her, acknowledging her crucial role in the appearance of the double helix.

How does the DNA Code Work?

The nucleotide bases make up the code of the DNA but how does this work to produce proteins? In 1961 Crick and Sydney Brenner showed that the genetic code must be read in triplicates of the bases, and these were called codons[6]. Later, in 1966, Marshall Warren Nirenberg cracked the code and deduced which codon codes for which amino acid for which he was given the Nobel Prize in Physiology and Medicine[7].

Let us step back a bit now. Proteins can be considered the workhorses of the cells as they carry out most of the basic functions of the cells, often after some modifications of their structure. These proteins are coded for by the DNA so that when a cell divides and becomes two cells, these daughter cells have the potential to make the same set of proteins as the parental cell, as the DNA within the daughter cells is an exact copy of the parental DNA.

Proteins are made up of a string of smaller molecules known as amino acids and it is the amino acids that the triplicates of nucleotide bases, the codons, code for. There are 20 amino acids that are used to make up proteins, and it is the order in which these amino acids are added that is important for structure and function of the protein. Combining the four nucleotides into three-letter codes results in 64 possible combinations, plenty to code for the 20 amino acids! Each amino acid is coded for by more than one codon except for one amino acid known as methionine which is always found at the start of a protein, as well as being found elsewhere in proteins. Three codons also code for what are known as STOP codons and these signify the end of a protein, that is they tell the cells to stop adding amino acids to the string.

However, the DNA does not code directly for the amino acids, there is an intermediate, just like a Head Teacher may instruct the classroom teacher what to teach the pupils. Here the Head Teacher is DNA, and the class-room teachers are analogous to RNA, the intermediate known as ribose nucleic acid (RNA). You can see that the only difference between the two names is deoxy, and this is because RNA contains a sugar known as ribose and DNA a sugar known as deoxyribose, that is deoxyribose has less oxygen atoms in its structure than ribose. There are two other important differences: RNA is single stranded, and instead of the T base, RNA contains the base U (stands for uracil). To make a new protein, firstly the DNA acts as a template for complementary base pairing to make an RNA, that is where there was a C in the DNA then there will be a G in the RNA, where there was an A there will be a U, a C there will be a G and a T there will be an A. This process is known as *transcription,* see Figure 1.3. This RNA is called messenger RNA (mRNA) as it carries the code for the protein. The mRNA then moves out of the nucleus into the cytoplasm where a specialised organelle (analogous to a small factory) is found, called a ribosome. This organelle is also made up of another RNA (ribosomal RNA) and proteins. The mRNA enters the ribosome where yet another type of RNA, called transfer RNA (tRNA), is found. Each tRNA can read the code in the mRNA for a particular amino acid and can add the correct amino acid. This process is known as *translation.* Thus, it is the order and the total number of amino acids that determines the type and size of the protein produced.

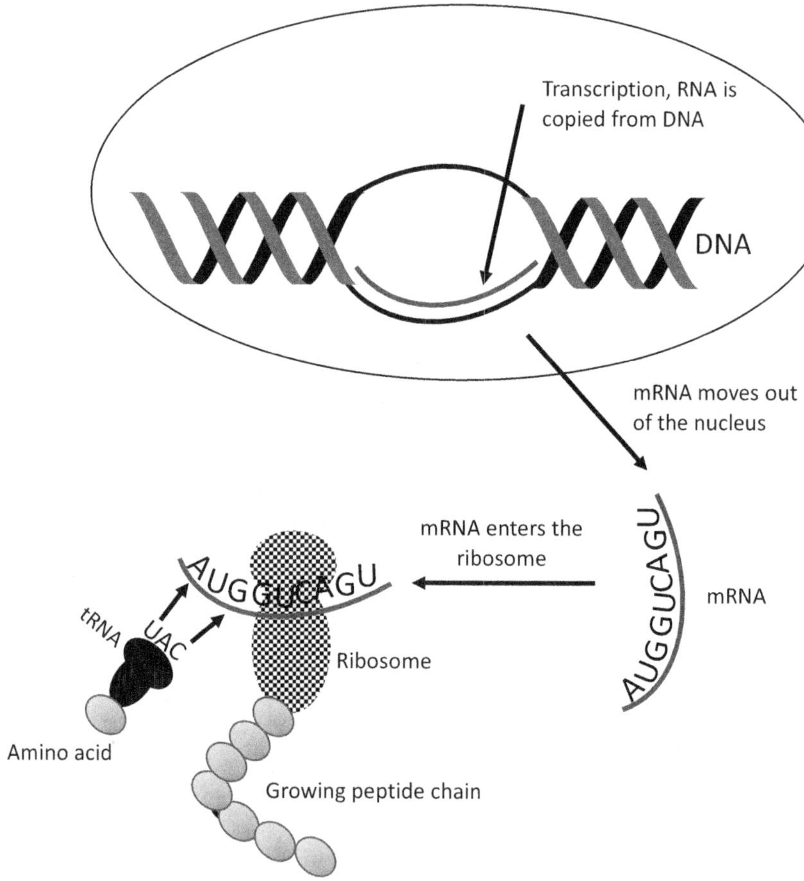

Figure 1.3: Proteins are synthesized using the DNA template of 3 bases via RNA. *In the nucleus the relevant gene is copied into RNA, G always copied to C, C always copied to G, A always copied to U (in RNA T is replaced by U), T always copied to A. The messenger RNA (mRNA) moves out of the nucleus into the ribosome factory present in the cytoplasm where transfer RNAs (tRNAs) read*

the bases to add the correct amino acid. Each tRNA transports a single amino acid to the ribosome and has a sequence that is complementary to that coding for the amino acid in the mRNA allowing it to "read" the code.

A very special mRNA

Most of us in the developed nations have recently become personally acquainted with a particular mRNA as the component of vaccines protecting us against Covid-19. This mRNA codes for part of the spike (S) protein (a protein that projects from the surface of the virus, like a "spike") in the virus coat and after injection into our muscle, the cells use the instructions in the mRNA to make part of the S protein. This causes the body to make antibodies and alert other immune cells which, upon infection with Covid-19, will act to eliminate the virus. In recognition for their work on the development of RNA vaccines, which made the development of vaccines like Pfizer-BioNTech Covid vaccine possible, the 2023 Nobel Prize for Medicine was awarded to Drs Katalin Kariko and Drew Weismann[8]. They found a way to make the mRNA acceptable to the host, by making small modifications which made sure that the RNA was not degraded and there were no unwanted inflammatory reactions.

The vaccine is estimated to have saved over 12million lives in the first year. However, the development of mRNA vaccines was not easy. Dr Kariko started working on mRNA vaccines in 1989 and from 1997 she continued the work in collaboration with Dr Weismann. Indeed, they faced years of scepticism and lack of funding. However, their perseverance certainly paid off and their work has great potential for application to other biomedical areas.

How is a large amount of DNA squashed into a small cell?

At this point we need to go into a little more detail about how DNA is packaged in the nucleus. The DNA in each cell is about 2metres long! As a cell is roughly around 10micro metres (μM) in diameter (this is approximate as cells differ in size) and there are 1000000μM in a metre, the DNA needs packaging! The DNA is packaged into bundles and each bundle is called a chromosome, see Figure 1.4.

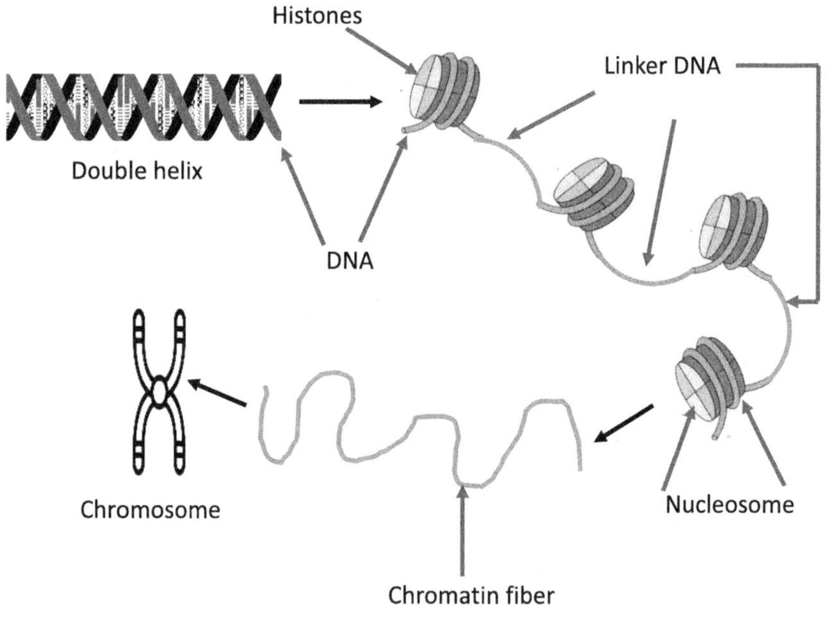

Figure 1.4: DNA is packaged into chromatin which form chromosomes. *The DNA double helix is wrapped around histones to allow packaging of the 2meters of DNA, rather like winding a hose pipe around a reel. DNA wound round histones is called a nucleosome and the nucleosomes with pieces of linking DNA form the chromatin fiber. These fibres form the chromosomes.*

In humans there are 23 pairs of chromosomes with each one of the pair coming from either our father or mother. When the cell is dividing and the nucleus breaks down, it is quite easy to see the chromosomes under the microscope, where they look like X shaped structures, see Figure 1.4. The chromosomes are labelled 1-22 with the sex chromosome pair being slightly different. These are called the X and Y chromosomes. When the sperm fuses with the egg at fertilisation we inherit one of each pair of chromosomes from our mother and one from our father. However, we always inherit an X chromosome from our mother and either an X or Y from our father. If the fertilised egg contains an X and a Y chromosome then the resulting baby will be a boy, if it contains two Xs (XX) then the baby will be a girl.

Okay, but being in chromosome bundles does not totally explain how this amount of DNA can be packaged into the nucleus. As was mentioned earlier, when nuclein was first isolated from cells in 1871 it was found to be composed of DNA and protein. We now know that DNA is packaged by being wound around proteins known as histones and the resulting DNA/histone complex is called chromatin. Figure 1.4 shows the folding of DNA from a double helix into chromosomes. There are different types of histones and H2A, H2B, H3 and H4 are associated with DNA. These proteins are positively charged and as DNA is negatively charged, the DNA can wind around the histones and bind tightly. The chromatin is made up of repeating structures called nucleosomes which consist of 146 base pairs of DNA and eight histone proteins. These nucleosomes were first observed by the husband-and-wife research team, Don and Ada Olins working at the University of Tennessee, and they described them as "beads on a string". They isolated DNA from rat thymus, rat liver and from chicken erythrocytes (unlike mammalian erythrocytes, bird red blood cells do contain DNA) using a series of new preparation methods and visualised it under the electron microscope. Using this powerful microscope, the nucleosomes could be clearly observed, and their research was published in the prestigious journal Science[9].

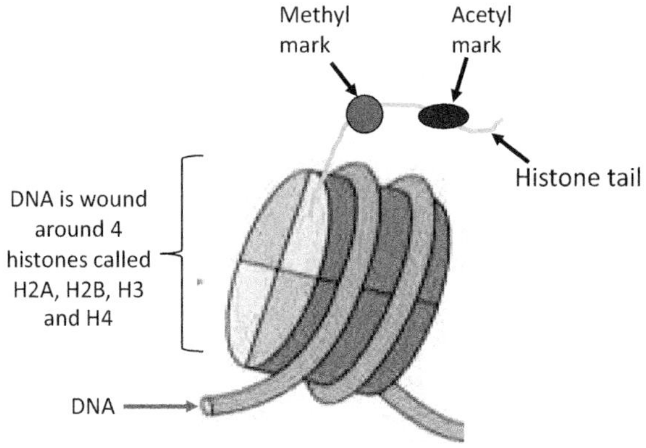

Figure 1.5: Chromatin structure with nucleosome. *DNA is packaged into chromatin which forms the densely packed chromosomes, visible at cell division. The DNA double helix is wrapped around histones to allow packaging of the 2meters of DNA, rather like winding a hose pipe around a reel. DNA wound round histones is called a nucleosome. The nucleosomes are continuously connected with stretches of linking DNA to form the chromatin fibers which can form the chromosomes.*

We now know that nucleosomes have a very distinct structure. Two of each of the four histones H2B, H2B, H3 and H4 form an octamer which binds 146 base pairs of DNA. Between each of these "beads" is a piece of DNA known as linker DNA (see Figure 1.4) which on average is 20 base pairs long. As, on average each chromosome contains over a 100million base pairs of DNA each chromosome must contain hundreds and thousands of these structures. Figure 1.5 shows a diagrammatic representation of a nucleosome. Interestingly, only organisms whose cells contain a nucleus with a membrane surrounding it have nucleosomes, single cell organisms like bacteria do not.

How Similar are We to Chimpanzees?

Our DNA is very unique to the individual and indeed this forms the basis for DNA identification that has become such an important tool in forensic science. In 1984 Sir Alex Jefferies discovered that within our DNA there are sequences that are repeated many times, these are called minisatellites and are short repetitive sequences usually about 10-60 base pairs long and occur at more than 1000 locations throughout the genome. Sir Alex discovered that these sequences were different among individuals and by comparing the DNA repetitive sequences it was possible to identify an individual. This became the basis of DNA finger printing[10]. Moreover, these sequences are inherited some from your mother and some from your father. At this time the minisatellite sequences could be identified as dark bands of different sizes on an X-ray film. This was achieved by using a radioactive probe. Thus, by comparing the films of the mother's and father's minisatellites, the parentage of an individual could also be determined. Sir Alex's discovery soon reached the newspapers and in 1985 he was approached by the lawyer of a Ghanaian lady who had been living in the UK, for a number of years, with her two children. She now wanted her third child, who had recently arrived in the UK, to stay with her but the Home Office believed his passport to be faked and he was in danger of being deported. Sir Alex took blood from the individuals, ran the gels and developed the film of the mother and all her children's minisatellite DNA and showed that the third child was the son of the mother and that all the children were indeed siblings.

It was not long before DNA finger printing entered the world of police forensic science. In November 1983, a 15-year-old girl was reported missing near Leicester and a few days later her body was found, just 300metres from her home. After a year of investigations, the police were no nearer finding the killer. In July 1986 the Detective Superintendent in charge of the case received a call to inform him that another young girl had gone missing. She too was found murdered in the same place as the first victim. A suspect came forward claiming he had murdered the second girl but not the first. Some of the details he relayed did not fit the crime scene, moreover the police were convinced that the same man murdered both girls but did not know how to prove this. The Detective Chief Superintendent thought of Alex Jeffrey's finger printing technique. He contacted him to see if he could carry out this technique on samples found on the two girls, which they believed belonged to the killer, and compare this to DNA from the person that confessed to the second murder. What Alex found was that the suspect did not kill the second girl as he claimed, but nor did he kill the first girl. However, the same person did kill both girls. This suspect became the first person to be proved innocent by DNA finger printing.

To try to find the killer, DNA was taken from every man aged 18-34 within 5 miles of the scene of the crimes. The budget for this had to be approved by the Home Office as it would cost over £500,000. Colin Pickford had asked a friend to give a sample instead of him, however the friend bragged about this in a local pub, and this was reported to the police. The police then arrested Colin Pickford and it was shown that his DNA was an exact match to that found on the two girls. Colin Pickford received two life sentences. Since then, DNA finger printing has revolutionised forensic science world-wide.

Well, how similar are we to chimpanzees? The minisatellite DNA that we have just discussed is a tiny part of our genome, and humans are 99.9% identical at the level at DNA. Moreover, we share over 98% of our DNA with chimpanzees and 60% of our DNA is the same as bananas!

Our protein-coding genes consist of stretches of DNA that code for a particular protein and many of these are so-called "housekeeping" genes that code for proteins that are involved in basic cell functions, and these are the same among mammals, insects and plants. Indeed, the fruit fly (Drosophila), is one of the most researched animals as nearly 75% of genes associated with diseases in humans are also found in this tiny fly.

There are around 20,000-25,000 protein coding genes in the human genome. This came as a bit of a shock in the year 2000, when the Human Genome project provided the complete sequence of a human genome, and it was found that protein-coding genes only made up around 2% of the DNA. The rest of the DNA, then known as non-coding DNA, must be doing something else although until very recently this was thought of as "junk". However, this certainly is not the case and much of this DNA is transcribed into RNA. Now genes can be divided into protein coding genes and non-protein coding genes (see next Chapter).

What makes that 0.1% difference between you and your neighbour?

If you look around at individuals on a bus or in the cinema or theatre, or anywhere where a large number of people can be seen, the first thing that strikes you is the difference between individuals. There is a range of heights, hair colour, skin colour, eye colour, facial features, weight and many other differences. Many of these, but not all, are determined by our genes.

Most of the differences between the DNA of individuals are the result of single base changes (known as single nucleotide polymorphisms, SNPs), or insertion and deletions of stretches of DNA. The latter two, insertions or deletions, being responsible for DNA finger printing used to identify individuals. However, in some cases these differences between individuals can have devastating effects. In 1984 Sir David Smithers, based at the Royal Free Cancer

Hospital in London, published a paper looking at the family trees of more than 450 breast cancer patients. He showed that breast cancer could occasionally run in families, with several relatives being affected by the disease across the generations. In 1990 there was a publication from an American group headed by Mary Claire King that a gene responsible for some of the hereditary observed breast cancers was carried on a particular chromosome named chromosome 17[11]. The actual gene was identified by a company known as Myriad Genetics and called Breast Cancer 1 (*BRCA1*[12]). The second gene, responsible for some other hereditary breast and ovarian cancers, was subsequently identified by a group based at the Institute of Cancer Research in Sutton, Surrey and called *BRCA2*[13].

Myriad Genetics caused a controversy in the world of biology, ethics and patenting by patenting the sequence of the genes coding for *BRCA1* and *BRCA2*. This was very disturbing - could part of the genome belonging to an individual be patented for commercial exploitation? Doesn't the genome of a person belong to that person rather than to a company? Indeed, in a landmark decision in June 2013, the US Supreme Court determined that DNA in its natural form cannot be patented.

In families carrying mutations in *BRCA1* and *BRCA2*, these mutations can be found in the genome of every cell in the body, including the sperm or the egg and as such are known as germ-line mutations. Having a mutation in either of these genes leads to an increased risk of breast and ovarian cancer. Mutations in *BRCA1* leads to a lifetime risk of developing breast cancer of 60-90% and of developing ovarian cancer of 44%. The chances of developing these cancers when a mutation in *BRCA2* is present is a little less. The risk of developing breast or ovarian cancer when carrying a mutation in *BRCA1* was highlighted by the actor Angelina Jolie who went public with her decision to undergo bilateral risk-reducing mastectomy as she is carrying a mutation in *BRCA1*. However, it should be noted that only about 10% of breast cancers are hereditary and only half of these are due to mutations in *BRCA1* or *2*.

DNA is a very dynamic molecule, and it has been estimated that an individual cell can undergo 1 million DNA sequence changes, (mutations), per day!! These can be caused by environmental factors such as UV light or replication errors during the copying of the DNA. Luckily the cell contains a machinery of proteins designed to correct most of these mutations and one of the functions of the BRCA1 and 2 proteins, that is disturbed when these genes are mutated, is involvement in the DNA repair pathway.

Back to how the DNA code works

Now we know more about our DNA, at this point we should go back to transcription and translation to get a little more detail. As described above, transcription is the process of how messenger RNA (mRNA), is formed. But we know that every cell contains the same DNA, so what makes a stomach cell function as a stomach cell or a liver cell as a liver cell? It depends on which genes are expressed, and this is regulated by ***Epigenetics.*** The mechanisms involved in the epigenetic control of gene expression will be discussed in the next chapter. Here it is sufficient to know that genes can be turned off, that is they are not expressed or turned on when they are expressed. How can this happen when the DNA is tightly bound around histones? It was originally thought that the organisation of DNA into chromatin, that is the binding of DNA to histones, was just a mechanism for packaging the DNA. However, it is now clear that chromatin restricts or allows the binding to DNA of factors that control the transcription of genes.

The expression of genes depends on the presence of specific modifications of the DNA itself and modifications of the histone tails. It can be seen from Figure1.5 that the histones the DNA is wound around have a tail sticking out and these tails can be modified by certain chemicals. One such modification can cause the tails to become more negatively charged which reduces the electrostatic interaction with the negatively charged DNA. The DNA itself can be modified with small chemical groups, and it is now clear that DNA modifications and histone modifications can "talk" to each other in contributing to gene expression. What we see is that when the chromatin is relaxed and more open due to histone modifications then transcription of the DNA is allowed as so-called transcription factors can bind. This type of loosely packed chromatin is called euchromatin. When the DNA is bound tightly to histone, DNA transcription is inhibited, and this type of chromatin is known as heterochromatin. However, binding of transcription factors does not always lead to gene expression as some factors have a repression domain which results in inhibition of gene transcription. Moreover, as in most aspects of biology, things are not as simple as this. There is a class of transcription factors known as pioneer transcription factors and these have the ability to bind to closed DNA, that is DNA that is tightly bound to histones. Once the pioneer factor has bound, it can displace a histone and/or recruit proteins that function to remodel the chromatin and so open the DNA ready for the binding of other transcription factors eventually resulting in gene activation.

A whole factory within the cell is required to result in the expression of genes. The transcription factors usually bind to a region of DNA that is upstream from the area of the DNA that is transcribed, and this is called the promoter region, as it promotes the expression of the gene. Enzymes known as RNA polymerases can now bind with the transcription factors. There are a number of different RNA polymerases in humans, and which is used is dependent upon which genes are expressed, but each RNA polymerase walks along one strand of the DNA known as the template. The RNA polymerase reads the code of one strand of the template DNA and builds a single strand of RNA through adding complementary bases, that is C is added where a G is, G where a C is, U (for uracil) is added where an A is and A is where a T is. In the case of protein coding genes, this messenger RNA now enters the cytoplasm and is translated into protein within the ribosomes as described above, see Figure 1.3.

However, yet another level of the control of gene expression is found within the cell. There are regions, other than promoters, within the DNA sequence that can activate transcription of genes. These can be located up to a million base pairs away from the gene to be transcribed and act independently of their orientation, distance, and location with respect to the target gene. These sequences of DNA are known as enhancers and are usually 100-1000 base pairs long. Transcription factors, chromatin modifying enzymes (see next chapter), other enzymes and a type of RNA polymerase known as RNAPII can bind to the enhancers and these can loop over to physically make contact with the promoter region. Enhancers can be found upstream or downstream of the target gene and even within the gene in the introns (see next section).

Genes are not coded for by a continuous piece of DNA

In multicellular organisms a single gene is not coded for by a single length of DNA but rather the gene sequence is interrupted by pieces of DNA that are not transcribed into mRNA. Parts of a gene that are transcribed into mRNA are called exons, and the parts that are not transcribed are called introns. The exons and introns are transcribed into preRNA but only the exons are spliced into mRNA, as shown in Figure 1.6.

Why do multicellular organisms have this arrangement? One reason is that the presence of introns allows for alternative splicing that means a single gene can generate several proteins. If you imagine that a single gene can be made up of four exons known as A, B, C, D interrupted by three introns known as a, b, and c. Thus, one protein could be coded for by exons A, B, C, D but a different protein can be coding for by exons A, C, D. As the sequence of amino acids determines the protein and the amino acid sequence is determined by the DNA sequence expressed as mRNA, then protein coded by A, B, C, D would be related to but different from protein coding by A, C, D, see Figure 1.6. Furthermore, recent research has shown that introns can be important in regulating gene expression.

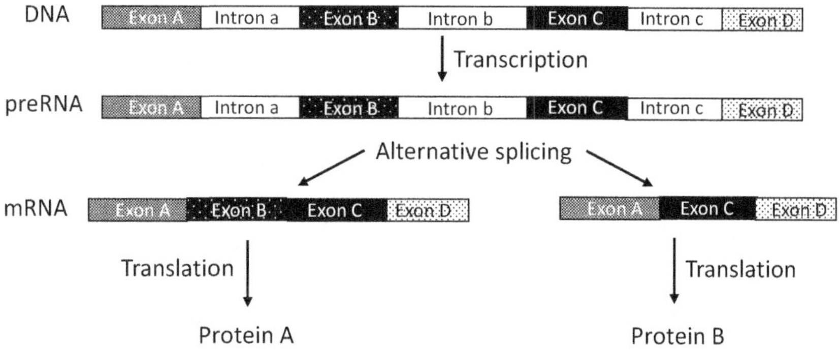

Figure 1.6: Genes are made up of exons and introns. *Exons and introns are transcribed into preRNA but only the exons are spliced into mRNA. Using different exons allows different proteins to be translated from a single gene.*

An Organelle with its own DNA

We have, in this chapter, talked about the DNA which is found in the nucleus (nDNA) on chromosomes, and this is the major constituent of our epigenome. There is however a set of organelles, found in the cytoplasm of our cells, the mitochondria, which are unique among the other organelles in the cytoplasm in that they have their own genome, mitochondrial DNA (mtDNA). The mitochondria have been called "the powerhouses of the cell" as they burn the food we eat to generate the energy we need for carrying out all of the cells' biological functions. There are many mitochondria in a cell, the number varying with different cell types, but all have multiple copies of their DNA. Their DNA codes for a few proteins (only 13) which are all involved in carrying out their unique function, and some non-protein coding RNAs. However, to perform their task they need help from genes in the nDNA that code for other proteins required for mitochondrial function. Moreover, changes in the nDNA such as the addition of small chemicals affect mitochondrial function, and mutations in the mtDNA (which mutates at a rate 100-fold higher than the nuclear DNA) can affect the nDNA and can result in debilitating diseases[14].

It is proposed that the mitochondria evolved around 1.45 billion years ago, and the cells of many animals, plants and fungi contain mitochondria. This organelle is thought to have originated from bacterial ancestors that entered the cell and stayed there, living in a state of "endosymbiosis"[15] as they developed into mitochondria. Supporting this proposed event is the fact that the structure of the mtDNA is very different from nDNA as mtDNA does not contain chromosomes or nucleosomes and resembles bacterial DNA in structure [16].

Unlike nuclear DNA, which is derived from a male and a female parent, mitochondrial DNA is unique in that whether you are male or female, this mtDNA is only inherited from your mother. This allows the ancestry of any individual to be tracked through their mother's lineage. and has given rise to the concept of "Mitochondrial Eve", and "The Seven Daughters of Eve", a book by Bryan Sykes[17]. By following non-pathogenic mutations which occur in mtDNA through evolutionary time, the people in Europe have been separated into seven tribes, developing from mitochondrial Eve who left Africa some time ago.

In this chapter we have seen that there are multiple factors and mechanisms for controlling nuclear gene expression, including DNA modifications, histone modifications, transcription factors and non-coding RNA. This multi-level and multi-mechanistic control emphasises how important it is to the organism to control gene expression. All our different cells can't be the same, even though they carry the same basic sequence of nDNA. Different tissues and organs perform different functions and to be able to do that, different genes must be expressed.

In the next chapter (Chapter 2), we will begin to explore the epigenetic mechanisms which are brought into play to control gene expression in order to maintain the functioning of the complex tissues and organs found in Homo sapiens (which make sure that teeth are not found in the liver)!! Having discussed these epigenetic mechanisms we will be in a good position in Chapter 3 to examine in more detail the structure and function of the mitochondrial DNA and the epigenetic mechanisms which have developed to integrate the actions of mtDNA and nDNA.

Glossary for Chapter 1

Chromatin	The basic unit of the genome containing DNA and histone proteins.
Chromosome	DNA is packaged into chromosomes which are visible using a microscope when a cell is dividing. Humans have 23 pairs of chromosomes, one of each pair from each parent.
Codon	This defines a sequence of 3 bases in RNA which defines the amino acid to be included in a protein.
Endosymbiosis	Two organisms living together with one inside the other, in this case the bacteria living within the cell becoming the mitochondria
Epigenetics	The study of how cells control the activity of their genes without changing the DNA sequence.
DNA	Deoxyribonucleic acid (DNA) contains all our genes which are defined by sequences of the four nucleotides, often called bases, Adenine (A), Thymine (T), Guanine (G), Cytosine (C). Through sugars attached to the bases, they are linked in a series and form the rungs of a ladder that hold the two strands of DNA together.
Dominant inheritance	A feature is inherited with only one copy of a gene.
Double helix	The two strands of nucleotides which make up DNA are twisted together in a double Helix where A pairs with T and G pairs with C.
Euchromatin	The loosely organised state of chromatin which allows transcription.
Exon	A sequence of DNA which is transcribed and is present in mRNA.

Gene	A sequence of one strand of DNA which is copied (transcribed) to form RNA which can either code for a protein as mRNA or can have a function of its own.
Heterochromatin	The tightly packed state of chromatin which inhibits transcription.
Inheritance	The appearance in offspring of features seen in the parents. (For example, the colour of peas in Mendel's experiments). Now known to be controlled by genes.
Intron	A sequence of a gene which is spliced out from preRNA and is not present in mRNA.
Meiosis	Action when a cell divides to produce two cells each of which has only one chromosome from each pair. This happens to produce sperm in males and eggs in females.
Messenger RNA (mRNA)	This RNA carries the code for the synthesis of a protein and is produced from preRNA by splicing.
Mini satellite	Short repetitive sequences 10-60 base pairs long which occur many times in the genome and are different for each individual.
Mitosis	This refers to the process of a cell dividing and producing 2 cells with the full complement of chromosomes.
Non-coding RNA (ncRNA)	RNA which is transcribed from DNA but does not code for a protein. Many ncRNAs have a function.
Nucleosomes	Repeating structures in chromatin consisting of 146 base pairs of DNA and eight histone proteins.
Pre RNA	The RNA which is transcribed from one strand of DNA which, after

	splicing out the introns, will give rise to mRNA.
Repressive inheritance	Two genes (one from each parent) are required for inheritance of a feature.
Ribosomal RNA	This RNA forms part of the structure of the ribosome.
Ribosome	A structure in the cytoplasm where proteins are synthesised.
RNA	Ribonucleic acid is single stranded and is defined by sequences of the nucleotides A, G, C, and Uracil (U) (which replaces the Thymine, used in DNA, and pairs with A).
Single nucleotide polymorphisms, SNPs	Variations in single nucleotides which are seen in individuals.
Splicing	The process of cutting out introns from preRNA to produce messenger RNA (mRNA) which codes for a protein.
Transcription	Term used to describe the expression of a gene to form RNA corresponding to the sequence of bases in one strand of DNA.
Transcription factors	Substances, usually proteins which can bind directly to chromatin to regulate gene expression.
Transfer RNA	A transfer RNA carries each amino acid to the ribosome as a protein is being synthesised.
Translation	In the ribosome, the code in a messenger RNA is translated to form a protein.
X and Y chromosomes	Females have two X chromosomes in a pair. Males have an X and Y forming a pair.
Zygote	The cell which is formed by fusion of a sperm with an egg to form a pluripotent cell.

Chapter 2
How is the Expression of Our Genes Controlled? Defining some Mechanisms of Epigenetics

"Unfading moths, immortal flies,
And the worm that never dies"
Rupert Brooke

As we saw in the last chapter, studies in the field of genetics in the 1950s showed that the genes which define our genome are found in DNA[1] and the double helix appeared to define the structure of nuclear DNA (nDNA). Knowing the structure of our genes, and how random mutations can give rise to changes in a "phenotype" advantageous for survival, fitted in well with the theory of Darwin and the development of new species. "Phenotype" defines the observable characteristics resulting from the controlled expression of genes. Also, the findings of Mendel took on a new dimension, where the traits in his peas could be defined at the level of the genome. However, defining genes and DNA at the molecular level inevitably led to other questions. Firstly, if all the (healthy) cells of our bodies have the same DNA with the same sequence, how can we develop cells and organs which have different phenotypes performing such different functions? Secondly, is the famous ability of Homo sapiens to adapt to his environment dependent only on random mutations of our genes, or can we, through epigenetic change stimulated by the environment, acquire a heritable phenotype more suitable to the environment? The first question relates to identifying the causal effects of how changes of the genome which do not affect DNA sequence modify phenotype, and this is the subject of this chapter. The second question relates to whether characteristics acquired in response to the environment are inherited in our progeny, and this we leave for later chapters.

Since the 1950s, studies in the field of Epigenetics have expanded dramatically and the emerging data have led to defining the field of Epigenetics as follows:

> ***Epigenetics:*** *Studies which examine the mechanisms where changes in chromatin which do not affect DNA sequence result in changes in phenotype and where the changes can be inherited*[1].

The studies also show that all the mechanisms which have been demonstrated to change the profile of the epigenome (and therefore the cell phenotype) involve the selective switching on or switching off of gene expression. Most of these studies have focused on nuclear DNA (nDNA) and among the multitude of factors which can affect gene expression, chemical modifications of DNA and histones play a major role. It is therefore appropriate to designate "chromatin" in the definition of Epigenetics since this is the functional unit of the nuclear genome which contains 99% of our DNA. As described in Chapter 1, chromatin is a composite of histone proteins and DNA, forming nucleosomes (see Figures 1.4 and 1.5 in Chapter 1) which can be loosely or tightly packaged thus influencing access for factors affecting transcription of the gene, allowing it to function. Factors affecting gene expression from the mitochondrial DNA, which is structured differently will be covered in Chapter 3.

Before embarking on describing the specific factors now known to be involved in regulation of gene expression, a look at the history of the development of the concept of Epigenetics provides interesting reading. It could be argued that we have a new date in the science of Molecular Biology, i.e. 1953, where we have: before the double helix (BDH) and after the double helix (ADH). While knowing the sequence of the nDNA of Homo sapiens has led to enormous advances in understanding what goes wrong in some diseases and in documenting individual differences between people, some scientists had remarkable vision BDH to see that Epigenetics was waiting in the wings to discover and explain so much more about the human condition.

Genotype to Phenotype: *Conrad Waddington, a Renaissance Biologist*

In the 1940s, before the molecular characterisation of nDNA, embryologists were asking the question - How can different phenotypes develop from cells carrying the same DNA? Life is dynamic and never more so than in the development of an animal from a fertilised egg where a single cell which is formed from the fusion of a sperm with an egg, produces all the different cells and organs of a new life. The cell arising from the fertilised egg is the zygote and the embryonic development of the zygote was largely studied with lower organisms where experiments could be done to begin to investigate how changes in genotype affected phenotype. These studies provided data which generated "innovative thought." Innovative thought triggers recognition of Conrad Waddington– an inspired individual who began to open the doors to finding an answer to this question and who first suggested the name Epigenetics to cover the studies trying to find an answer.

He first studied in Cambridge in the field of palaeontology, but he was a polymath, awarded a fellowship in moral philosophy as well as printing and editing a poetry magazine. However, he slowly gravitated to working in the field of embryology, where he could study and document the development of all the different phenotypes emerging from the zygote. Dame Honor Fell describes how she met him when he had his fellowship in moral philosophy, and he asked her if he could come to her laboratory to learn and adopt the techniques used there for studying embryological development. She did accept him in her lab at the Strangeways Research Institute in Cambridge and his career as an embryologist took off and led later to his being offered the Chair of Genetics in Edinburgh, where he built up the Genetics department to be the largest in the UK, and one of the largest in the world. He had been known to say that he was surprised when his ideas were taken seriously by colleagues when he was so young, but Waddington went completely bald at the age of 21, and perhaps he was taken to be older than he was[2]!

The development of his thought at this time (1940s) is recorded in his publications, where he suggested the word Epigenetics to cover the network involved in the pathway from genotype to phenotype[3,4]. The prefix epi is from the Greek, meaning above or beyond, so it refers to factors above or beyond the sequence of the genome. He was very influenced by the work of Spemann who established that different parts of the embryo could induce the differentiation pathways of adjacent cells and Waddington used the word "evocator" to describe the signal generated to induce these changes. His attempts to identify the evocators were unsuccessful. However, from the work of Thomas Hunt Morgan and colleagues with the fruit fly (Drosophila), the Chromosomal Theory of Inheritance was established, and it was shown that genes are located on chromosomes. Hunt Morgan was awarded the Nobel Prize for Medicine in 1933 for this work[5]. In thinking of how the evocators from the inducing embryonic tissue could affect the phenotype, Conrad Waddington believed that regulation of gene expression must be involved:

"In embryonic induction the cells must also contain some regulator substances capable of specifically activating or derepressing certain genes" and "An adaptive response to an environmental stimulus depends on a suitably genetically controlled reactivity in the organism".

Hunt Morgan, of Drosophila fame, had written a book in 1934 called, "Embryology and Genetics[6] and Waddington notes that Morgan "firmly advocated the point that the fundamental agents that bring about embryonic development are the genes and that the only satisfactory theory must be a theory of how the activity of genes are controlled".

At this time Waddington discusses the dominance of the Darwinian theory of evolution based on random mutations leading to changes in phenotype which improve survival, and to the general dismissal of the Lamarckian theory of inheritance, where characteristics acquired by interaction with the environment, are inherited. The dismissal of the theory of inheritance of acquired characteristics was attributed to lack of evidence but the appeal is still there as a possible solution to explaining the adaptation of organisms to the environment and Waddington suggested that the inheritance of acquired characteristics should be re-examined. Indeed, after the second world war, documentation of phenotypic data from human cohorts, involuntarily subjected to changes in nutrition, suggest that changes induced by the environment can be inherited by offspring. These data will be examined in Chapter 5. Here we will confine ourselves to outlining some of the molecular mechanisms underlying the changes in gene expression which lead to definition of cell phenotype, while having no effect on DNA sequence.

The Profile of Gene Expression Defines Cell Type (Phenotype)

As predicted by Waddington and Morgan, the changes in phenotype which occur in embryonic development depend on the activation or repression of specific genes. Certainly, regulation of gene expression is core to developmental pathways and some of the molecular mechanisms underlying this regulation have been defined.

Most of the studies identifying these mechanisms have been carried out in simple systems using cells cultured *in vitro* (outside of the organism; in the laboratory) and/or with lower organisms, such as Drosophila (flies) and amphibians (frogs, toads). The cultured cell lines were originally developed from cancer cells which continue growing given a nutritious medium, while explanted normal cells have a limited life span. Also, in the development of the embryo the zygote divides to produce cells which give rise to two separate structures: the placenta, which will feed the embryo, and the inner cell mass (ICM) which will give rise to all the different cells of the new organism. Cells of the ICM are pluripotent, that is they are stem cells that have the ability to give rise to all types of cells. They can be grown in culture as embryonic stem cells (ESCs) and with selected growth media can be induced to differentiate into the different cell phenotypes. As the cells acquire their functional phenotype, genes are switched on or off as the cell differentiates. Importantly, as the cells go down a particular pathway of proliferation and differentiation, the genes not required for their function are permanently switched off ensuring that teeth are not formed in the brain or nails in the liver. Figure 2.1 is a diagrammatic representation of Waddington's famous illustration showing the continuous differentiation occurring in the embryo. He depicted the zygote, as a ball rolling down a hill dividing into separate troughs where different cell types develop[7]. We have here represented the zygote as a ball rolling downwards on a bagatelle table, where the troughs are created by columns. These troughs represent separate differentiation pathways where they become terminally differentiated and functional, and cannot go back up the hill or change direction but stay in the trough to form a specific organ or tissue such as the liver. While this is true in the development of the embryo, it has now been possible to take the fully differentiated cell from an adult animal and force the backword path to the original multipotent cells. This is done by overexpressing certain genes as was evidenced by the cloning of the sheep Dolly.

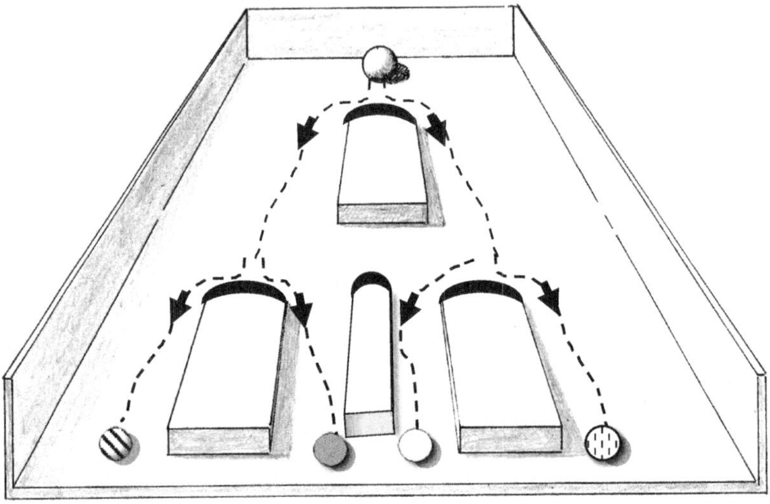

Figure 2.1: Representation by the use of a bagatelle table of Waddington's famous diagram showing part of the epigenetic landscape, which visualizes the development of the embryo: *The path followed by the ball, which represents the multipotent fertilised egg i.e. the zygote, as it rolls down the table, corresponds to the developmental history of the cells as they develop to form the different cell types (phenotypes) which will form the embryo. The different phenotypes are represented by spheres with a different filling from the multipotent zygote. In the diagram, the ball can initially roll to the left or the right, taking either of these routes leads to another differentiation pathway alternative. Multiple troughs and thresholds develop, here illustrated by the column shaped blocks, which stop the cells shown from veering from their course of differentiation. The downhill progress indicates that in the embryo the development cannot be reversed.*

After birth the animal has to maintain the dynamic balance of each tissue. In a healthy adult, while each tissue or organ does not expand in size or shape, the situation is highly dynamic and resembles embryonic development. Thus, stem cells specific for each tissue divide slowly to produce cells which divide more rapidly and then differentiate and - with time – die, to be replaced by new functioning differentiated cells. This is illustrated in Figure 2.2.

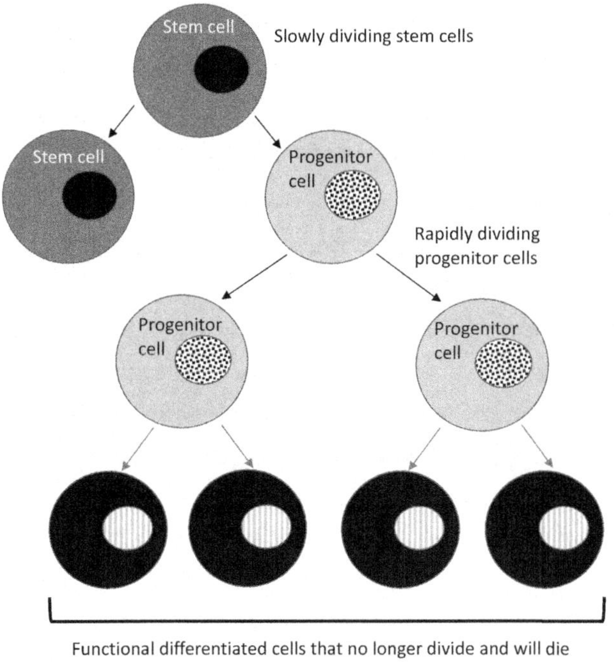

Figure 2.2: The dynamics of normal tissue development in the adult animal. *The slow-dividing stem cells in a tissue give rise to the rapidly dividing "progenitor" cells which in turn produce the differentiated functional cells. These cells function for a while and then die to be replaced by new functioning cells. While the DNA sequence stays constant, the epigenome changes and determines the profile of genes expressed.*

Importantly, at the stage of development where cells are dividing, the profile of the epigenome is stably carried into the daughter cells at cell division (termed mitosis). Thus, the situation in adult tissues is very dynamic in maintaining the balance between proliferation and differentiation in a healthy tissue. In cancer, which is a proliferative disease this delicate balance becomes unbalanced (see Chapter 4).

How is the expression of our genes controlled?

So, we have discussed that the cells making up different tissues express a different profile of genes but how does a cell know which genes to turn on and which ones to keep silent?

This is achieved in a number of different ways: Firstly, the transcription factors (see previous chapter) expressed by different cell types will influence which genes are expressed by those cells, but the expression of the transcription factors themselves has to be controlled! Secondly, the chromatin can be modified by the addition of small chemicals. Such modifications occur on the DNA within the chromatin and involve the presence or absence of methyl groups. In addition, the histones within the chromatin can also be modified by small chemical groups and most commonly this can be the addition or removal of an acetyl group or a methyl group (see Figure 1.5 in Chapter 1). Thirdly, and perhaps surprisingly, non-protein coding RNAs make up a very high proportion of the total RNA transcribed by the cell and although the function of many of these RNAs is still to be defined, the function of some is well-known. Small stretches of RNA known as micro RNAs (miRNA) are involved in the silencing of protein coding genes and long non-coding RNAs (IncRNA) can affect a wide spectrum of functions within the cell, including controlling the expression of genes. Thus, the many levels of gene control and the high volume of non-coding RNA expressed by cells testifies to the enormous importance of controlling gene expression in development and in maintaining the complex adult tissues of higher organisms and indeed humans. We can now go into more detail on some of these methods of transcriptional control.

Addition of Small Molecules to DNA and Histones can Control Gene Expression

Epigenetic modification of DNA

The first modification of DNA to be documented which affects phenotype is one which inhibits gene expression. This change involves adding a methyl group to one of the bases in DNA. The base is cytidine (C) and in the promoters of many genes, there are clusters of sequences where C is linked to a guanine (CpG). These are called CpG islands and methylation of cytidine in these islands results in reduction of gene expression, and in some cases the gene is switched off.

The laboratory of Professor Adrian Bird played a major role in the study of DNA methylation and the enzymes involved in adding the methyl group as well as identifying a "reader" which detects the methylation[8]. As Professor of Genetics in Edinburgh (and Director of the Wellcome Trust Centre for Cell Biology), Adrian Bird recalls how he first encountered DNA methylation while working as a post-doctoral fellow (1973-1974) in the laboratory of. Professor Birnstiel in Zurich. Professor Birnstiel was known for his work on ribosomes (the structures in the cytoplasm supporting the translation of RNA into protein, see Chapter 1), which was also Adrian's interest at the time. However, this was a time when scientists were working with and preparing "restriction" enzymes which recognised specific sequences of DNA where they could cut the DNA, and a fellow scientist, who was on sabbatical in the same lab (Hamilton Smith, known as Ham), was preparing these enzymes. Dr Bird had preparations of DNA coding for amphibian ribosomal RNA which carried methyl groups (on cytidines) and preparations which did not carry methyl groups. On testing one of Hamilton Smith's restriction enzymes for its ability to cut the two preparations, he found that the enzyme cut the non-methylated DNA but did not cut the DNA carrying methyl groups, (see Figure 2.3).

Importantly, Ed Southern who had just developed the technique for detection of the size of DNA (Southern blotting) was also in the Birnstiel lab on sabbatical, and the technique could be used to assay the activity of the restriction enzyme. Hamilton Smith later won the Nobel Prize in 1978 for his work on restriction enzymes, Ed Southern became Sir Edwin and the interaction of the three young innovative scientists led to this important step in the study of DNA methylation, enabling sequential sequencing of a piece of DNA to map methylation. In a later interview (2009) Professor Bird describes this scientific interaction in a modest way:

"*I'm not sure what I'm doing, to be honest, I wanted to do something interesting but I was just playing around more than anything else. Ham was playing around with the enzymes he knew about and Ed was doing the technical things he liked doing. So accidentally this gave rise to the idea that one could use restriction enzymes to map methylation in DNA*" (see Figure 2.3)[9].

> *There is a message here for young research scientists - interact with other investigators in the laboratory who are working on a different project!*

Bird and his co-workers then went on to map the distribution of methylated cytidine in the genomes of different species. They found that in the genome from vertebrates (includes us), a high proportion of the DNA is methylated but some small regions of DNA were not methylated. These regions contained unmethylated CpG islands and were located in the promoters of genes which were actively transcribing. There is now strong evidence that methylation of CpG islands in gene promoters acts to repress gene expression, that is the genes are switched off. Moreover, changes in the profiles of genes which are repressed are associated with some diseases (see below and Chapter 4 on cancer).

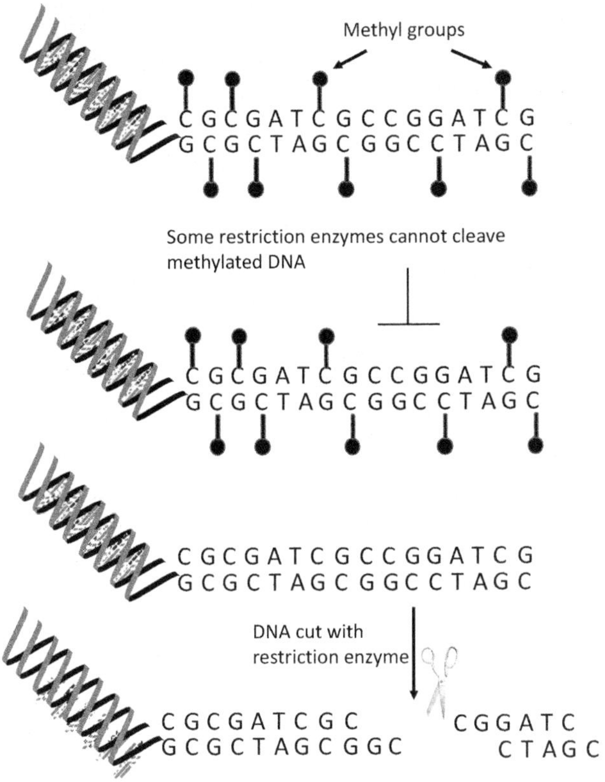

Figure 2.3: A method for identifying sites of methylation of DNA*. Restriction enzymes can cleave DNA at certain sequences, but some of these enzymes cannot cleave the DNA if the target sequence is methylated. In this way it has been possible to determine which sites in DNA have been methylated.*

The group then proceeded to search for a "reader" of the methylated cytidine which might lead to defining the mechanism of gene repression. Surprisingly the protein they identified which "reads" (i.e. binds to) the CpG methyl mark - MECP2 - was found to be mutated in a particularly debilitating neurodevelopmental disorder. This disorder is named the RETT syndrome, which develops mainly in girls. The gene coding for MECP2 is on the X chromosome and a mutation in one allele is found in these girls, who begin life normally after birth and then lose control of arm and hand movement and have other mobility and neural problems[10]. It is a terrible disease, seemingly healthy babies begin to deteriorate in how they can move, walk and talk.

There is variation in the time at which the girls begin to show motor abnormalities and in the severity of the debilitation. These differences can be attributed to the fact that girls have two X chromosomes, and in each tissue one X chromosome is "inactivated" (see below under *XIST:*X-inactive specific transcript*)* and the other "active" chromosome is expressed in a mosaic fashion. In this rather haphazard way of selecting the "active" X chromosome, the one carrying the mutated *MECP2* gene can be more or less dominant.

As yet there is no therapy for the girls with RETT syndrome, but extensive research is being undertaken to understand what is happening at the molecular level in the hope of identifying a possible therapy. Families with children with RETT syndrome have interacted internationally to establish a charity supporting this research. The *MECP2* gene is only found in vertebrates so that in working with animal models, it is not possible to use lower organisms like flies and amphibians, and so the research is being conducted in mice which do have a *Mecp2* gene[11]. *Mecp2*-mutant mouse models recapitulate a broad spectrum of phenotypes seen in human RTT patients and targets have been identified which suggest the possibility of therapy. These studies provide hope that as technology advances, children with this terrible disorder may be helped. The manifestation of the RETT syndrome - though awful - does demonstrate how important the correct reading of the CpG methylation mark in DNA is. The importance of this mark in another less rare disease, cancer, is discussed in Chapter 4, where we describe the epigenetic changes occurring in cancer, and the development of inhibitors of DNA methylation for use in cancer therapy.

Epigenetic modification of histone proteins

As we saw in Chapter 1, the double helix of nDNA is wrapped around histone proteins forming nucleosomes along the chain, and the whole structure is known as chromatin (see Figure 1.4 in Chapter 1). Regions where packaging of nucleosomes in chromatin is tight means that accessibility of factors to DNA that can regulate gene expression is very limited. Where the packaging is looser, the DNA in the chromatin can be accessible to these factors. Importantly, epigenetic modification of the histone proteins can play a major role in regulating the openness of chromatin structure. This was first observed and characterised by Vincent Allfrey in the early 1960s[12]. He was certainly a visionary and began to look into the histone modifications and to investigate their possible effects on transcription. He foresaw that:

"minor modifications of histone structure, taking place on the intact protein molecule offer a means of switching on or off RNA synthesis at different loci along the chromosome"

Until then the histones were just seen as "glue" holding the chromatin together and a way of packaging DNA. Most scientist feel that it is completely appropriate that Vince Allfrey should be singled out as being one of the forefathers of epigenetics and it is sad that he died in 2002 and so missed seeing the field of epigenetics explode.

The histone proteins have "tails" which are not wrapped by the DNA and in these tails are amino acids which can be enzymically modified by adding a small chemical group. The most important epigenetic changes occur to the amino acid lysine (K) and two important changes are the addition or removal of acetyl or methyl groups on lysine (see Figure 1.5 Chapter 1). Amino acids can be represented by a letter, and K is used for lysine.

Acetylation of histones

An acetyl group is a small chemical made up of two carbons, an oxygen and three hydrogen atoms and it is now clear that acetylation of histones can activate gene expression. It is relatively easy to see why adding an acetyl group increases the looseness of chromatin. Acetylation changes the overall charge of the histone from positive to neutral, and so the interaction of the histones with the negatively charged DNA is reduced and therefore condensed chromatin becomes more relaxed and open. Thus, by increasing the accessibility of hidden DNA target sites, histone acetylation allows proteins required for transcription to occupy the relevant gene promoters[13]. There are enzymes which can add the acetyl groups, and these are called histone acetyl transferases (HATs) and this modification can be removed by enzymes which take off the acetyl groups, i.e. by histone deacetylases (HDACs). We will see later how inhibitors of HDACs have now been incorporated into the arsenal of cancer therapy in Chapter 4.

Methylation of histones

Although at the time they were not directly working on histone methylation, scientists relating genetic mutations to changes in phenotype in the fruit fly did lay the foundations for the study of histone methylation by identifying two sets of protein complexes, coded for by the so-called polycomb and trithorax gene families that have opposing functions[14]. The exotic names that the fly scientists gave the genes relate to observed phenotypes in the mutated flies. Polycomb refers to the bristles which grew where they should not, and thorax refers to changes that were observed in the body structure. These genes are conserved in evolution and the polycomb and trithorax proteins act antagonistically with each other to control gene expression. It became clear that each group of proteins acted as complexes containing other proteins and much later it was found that the complexes contain methylases which could add methyl groups to some lysine amino acids found in histones. Moreover, these enzymes were very selective as to which lysine they methylated. The polycomb complexes attach methyl groups to lysine in amino acid position 27 of the tail of histone 3 (H3K27) and in doing so act to *repress* gene transcription, see Figure 2.4.

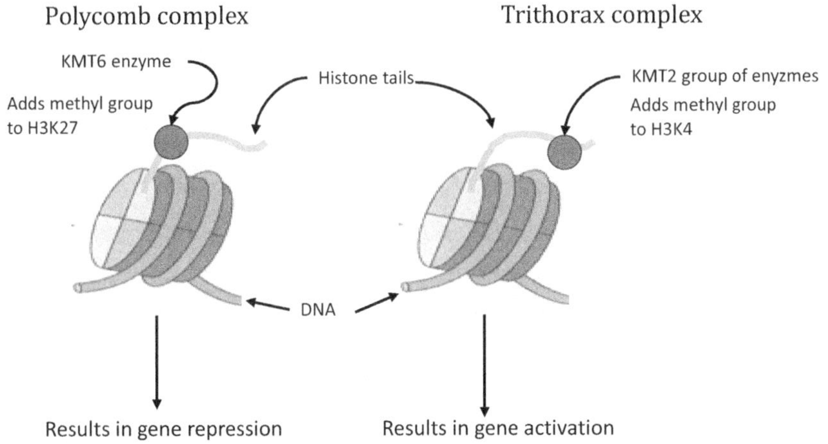

Figure 2.4: Two complexes known as polycomb and trithorax act antagonistically to contribute to controlling gene expression. *An enzyme called KMT6 in the polycomb complex adds methyl groups to histone 3 lysine 27 (H3K27) and the addition of this methyl*

groups inhibits gene expression. The KMT2 enzyme in the trithorax complex adds a methyl group to histone 3 lysine 4 (H3K4) and the addition of this methyl group results in gene activation.

Thus, the enzymes which add methyl groups to lysines are called KMTs, the polycomb enzyme being KMT6 (alias EZH2 in Chapter 4). On the other hand, trithorax complexes, having different methylases, add methyl groups to lysine in position 4 of the histone 3 tail (H3K4), and these methylases (KMT2 group) *activate* transcription, see Fig. 2.4. The polycomb and trithorax complexes function at gene promoters so can act antagonistically to regulate gene expression.

For several years it was thought that histone methylation was irreversible, but in 2004 a histone demethylase which takes off methyl groups from H3K4me2 was identified (KDM1) and now a whole panoply of histone demethylases (KDMs) have been identified - again showing specificity for the lysine carrying the methyl group. There are other lysines in the histone tails where other specific methylases and demethylases can act in repressing or activating gene expression. Therefore, the methylation state of the epigenome of a cell, is complex and contributes significantly to defining its phenotype[15]. Since some of the lysines can be either methylated or acetylated, this adds another dimension to the modification of histone tails and so to the profile of the epigenome. In most cancers this profile is dramatically altered and therapies based on targeting the enzymes adding or removing the acetyl or methyl mark are already in the clinic for cancer therapy (see Chapter 4).

The New World of RNA Transcribed from Genes not Coding for Proteins

Until now we have mostly discussed genes as though they were all coding for proteins. The rest of the genome, which does not code for proteins was thought of as "junk" DNA, but it is now clear that there are other genes which code for RNAs which do not code for proteins, but which do have a function themselves. Scientists have been aware for many years, of the existence of a few specific functional RNAs such as ribosomal RNA, which forms part of the structure of the ribosomes, and transfer RNAs, which take the amino acids to the ribosomes for protein synthesis (see Chapter 1, Figure 1.3). However, as techniques in genetics have moved forward, a whole new world of functional RNA has emerged, opening-up new dimensions in the control of gene expression and cell fate.

Following the success of the "Human Genome Project", where a large International Consortium worked together for several years to determine the sequence of the human genome, an expanded international consortium was assembled to define the factors involved in gene expression and gene function. This is "The Encyclopaedia of DNA Elements" the ENCODE project, and their findings are now published 30 years after the initiation of the Human Genome Project. Reflecting the large number of individuals and institutes involved in ENCODE, the number of authors listed on the publications is huge. The staggering result of their work is the finding that 75% of the human genome can be transcribed to produce RNA[16], showing that this is certainly not "junk" DNA.

Amongst themselves, over the years they were working together, the scientists in ENCODE organised a sweep stake to guess how many protein coding genes were in the human genome. Most estimated a large number like 100,000 or more with only one or two guessing less than 30,000. The high estimate assumed that with the tissue complexity seen in humans (particularly brain) there must be a large number of protein coding genes. It turns out however that only around 20,000 protein coding genes are in the human genome, i.e. around 2% of the genome and similar to the number found in lower organisms. This suggests that there may be multiple genes coding for functional RNA and that these could play an important role in the epigenetic control of gene expression and cell fate in complex multicellular organisms. Sorting out which of these non-protein coding RNAs do have a function will be ongoing for some time. However, where functional RNAs have been identified and characterised, they have been found to play multiple roles in cell metabolism and gene regulation. Moreover, some of them are aberrantly expressed in many diseases including cancer.

While it is early days to have a definition of function for all of the newly identified RNAs, we can discuss 1. the "micro RNAs" (miRNA) whose function has been well defined, and 2. some examples of the "long non-coding RNAs" (lncRNA) where enough information regarding their function is available to make them of great interest. 3. The function of a group of non-coding RNAs (ncRNAs) which do not leave the nucleus but define territories there involved in gene regulation.

The micro RNAs

The story of finding and identifying micro RNAs begins with studying a worm, Caenorhabditis elegans, or C elegans. When the double helix appeared, a young Oxford graduate student (Sydney Brenner) visited Frances Crick before the announcement of the structure of DNA was made public. He decided then that he needed to find a model organism where he could map and study all the genes and work out their functions. He said:

"We could now find out how genes worked. On that day I knew exactly what I wanted to do"

Drosophila, the fruit fly which came to fame in the Morgan lab was too complex and Sydney Brenner, then 26 years old, wanted something simpler and so he chose the worm C. elegans, with its six chromosomes, with the aim of identifying each individual cell and looking for the functions of all the genes. He had to wait until he was established to obtain the financial support he needed but he did manage to start work on the genes of this worm in 1967 and published his findings in the early seventies. There is a famous extract from a one-page grant application submitted to the MRC (Medical Research Council):

"To start with we propose to identify every cell in the worm and trace lineages. We shall also investigate the constancy of development and study its control by looking for mutants".

Before obtaining the grants however, after the meeting with Frances Crick, a collaboration was established with him to sort out the details of transcription and translation and Brenner published with Crick the first paper identifying messenger RNA in 1961[17]. The model organism selected by Sydney Brenner has provided an invaluable tool for exploring the basic molecular mechanisms of life and has been used globally by multiple investigators. Indeed, it is in working with C elegans that the first microRNA was identified. Which of us could have realised that worms could provide such an important biological tool! Sydney Brenner knew exactly what he was doing with his selection, and when he was later awarded the Nobel prize, it was not only for his own brilliant work but also for his influence on the progress of molecular biology globally.

So, it was a group following the development of the larvae of C. elegans who first encountered microRNA. For this important discovery of microRNA Victor Ambros and Gary Ruvkun were awarded the Nobel Prize for Physiology or Medicine in 2024. They found that a gene (called *lin4*) when expressed, could control the level of a protein coded by the gene *lin14* which is required at an early stage of development, but is down regulated at a later stage. They also knew that the *lin4* gene did not code for a protein, so how could *lin4* act to repress expression of the LIN14 protein? Examining the sequence of the *lin4* RNA (only 22 nucleotides long) they found it had a sequence that was antisense to sequences present in the messenger RNA (mRNA) coding for the LIN14 protein[18]. As illustrated in Figure 2.5, this means that where there was a C in the mRNA, there was a G in the *lin4* sequence and where there was an A in the mRNA there was a U in *lin4*. Thus, this anti-sense sequence could block translation of the mRNA into LIN14 protein by binding to its complementary bases, and a new pathway for regulating the function of protein coding genes was uncovered.

The search for *lin4* miRNA in vertebrates was unsuccessful, but 7 years later another miRNA was identified in C elegans - *Let 7* and this miRNA was found across the evolutionary spectrum, and in humans there are several forms of *Let 7*[19]. It has now been found that the *Let 7* miRNA is a tumour suppressor, able to down regulate expression of a gene whose expression can promote tumorigenesis. *Let 7* is down regulated in several cancers, particularly lung cancer,

Thus, each microRNA (miRNA) uses a completely new method of inhibiting the function of a protein coding gene by blocking the function of its specific messenger RNA, and so inhibiting protein production. Although the first documented miRNA blocked translation of the target mRNA, the antisense binding of the miRNA can block the mRNA function in different ways, for example by decreasing the stability of the mRNA.

The genes for miRNAs are ancient genes, being found in lower organisms, including C elegans where they were first detected. MicroRNAs not only provide a novel epigenetic route for regulating protein expression, but a technology has been developed based on miRNA where expression of a specific protein can be inhibited. This is done by delivering a small RNA (shRNA) to the cell with a sequence which is antisense to the RNA coding for the protein of interest. By being able to block the expression of a protein at will, its function can be readily investigated. More recently a new engineering tool has been developed by scientists to edit DNA. This is called CRISPR and it can be used to delete or correct precise regions of DNA. Efforts are ongoing to investigate its usefulness in treating some diseases.

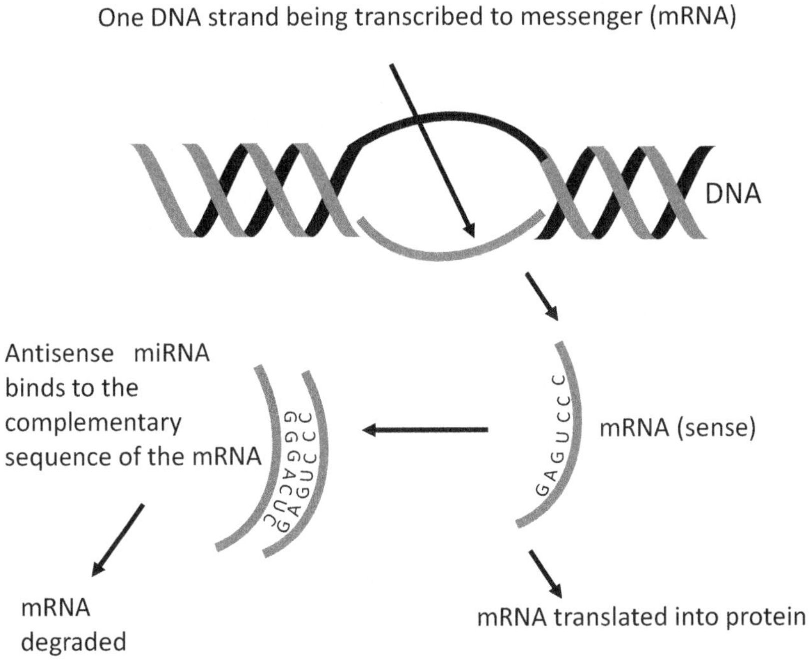

Figure 2.5: A messenger RNA (mRNA) can be translated into a protein, but the mRNA is degraded if it is bound by an antisense microRNA (miRNA). *Here, a small part of the LIN14 mRNA is shown bound to lin4 antisense miRNA (see text) and is then degraded.*

There are around 2000 miRNAs in Homo sapiens and the profile of their expression changes during development and differentiation, thus modifying the epigenome and cell fate. The miRNAs are produced at high levels compared to messenger RNA (mRNA) and can be found in serum and circulating white blood cells. Taking blood is considered a "non-invasive" technique and although measuring the spectrum of miRNA in serum does not necessarily inform on the source of the miRNA, experiments are already underway to measure the profiles of circulating miRNA in cancer patients, and in looking for responses to varied nutrition or smoking in specific cohorts of individuals. These studies will be discussed in Chapters 4 and 6.

Long non-coding RNAs (lncRNA)

Among the expressed RNAs which do not code for proteins (non-coding or ncRNAs), the ***long*** non-coding RNAs (lncRNAs) have been classified as having a length of more than 200 nucleotides, and the number of lncRNAs with a defined function is continually expanding. Unlike the miRNAs where each of which operates with the same basic function, which has been well defined, each lncRNA is functionally unique, and together they affect a wide range of cellular processes. These include chromatin structure and the modification of the epigenetic marks discussed above. Moreover, since changes in expression of these RNAs are seen in many diseases, each individual lncRNA is often studied in the context of a particular disease. To give an idea of the new vistas being opened by lncRNAs and to illustrate their diversity of function, we can discuss the functions of three lncRNAs which have been widely studied, namely *XIST*, *HOTAIR* and *MORT*.

XIST (X-inactive specific transcript)

The *XIST* lncRNA and its involvement in inactivating one of the two X chromosomes in the development of female embryos was studied long before the findings of the ENCODE project showed that 75% of the genome is transcribed and the class of lncRNAs was defined[20]. Mammals which have a placenta, which covers most mammals except marsupials and the egg laying ones, appear to be the only animal group which absolutely require the contribution of DNA from a female and a male to make a viable embryo and *XIST* RNA is expressed in all mammals operating with a placenta to feed the developing embryo.

Perhaps we should go back a bit and remind ourselves that one of the two X chromosomes in the developing female embryo comes from the father (who has the XY pair of chromosomes) and the other from the mother, who has the XX chromosome pair. Thus, all the female eggs have an X chromosome. On the other hand, each sperm can be carrying the Y chromosome to produce male embryos or an X chromosome to produce female embryos. The small Y chromosome has very few genes and these are related to developing the male phenotype, but the X chromosome has over a thousand genes which affect many facets of metabolism. Therefore, there would be an imbalance of gene expression between males and females if the genes on both X chromosomes were expressed in the female and only one in the male, so this needs to be fixed. This is done by blocking the function of the genes on one of the X chromosomes in the female embryo and XIST is crucially involved in this inactivation. The choice as to which X chromosome is to be inactivated is random and as the differentiation of the embryo proceeds, the pattern of expression of the active X chromosome is mosaic- with some of the cells expressing the maternal X and some expressing the paternal X. As we saw above in the discussion of girls with RETT syndrome the choice of which of the X chromosomes is selected to be the active X chromosome can impact on the severity and timing of the disabilities in RETT syndrome sufferers.

The XIST protein covers the X chromosome to be inactivated and recruits factors such as histone deacetylases which repress gene expression. However, although most of the genes on the "inactivated" X chromosome are not expressed, around 20% of genes on this chromosome are not silenced and these genes are expressed by both the active and "inactive" X chromosome in females. This means that if a mutation occurs in one of the genes that are not silenced, there is a back-up from the non-mutated gene found on the other X chromosome in women. On the other hand, mutation of one of these genes on the only X chromosome found in males can give rise to what have been called X linked diseases in males where there is no back up as there is in females. This has been documented for mutations in one of these un-silenced X chromosomal genes (an H3K4 demethylase, *KDM5C*) where men with such a mutation show defects in neuronal development[21].

HOTAIR (HOX transcript antisense RNA)

In order to begin to talk about *HOTAIR* it is important to discuss the site in the genome where the gene coding for *HOTAIR* lies and the genes it finds itself surrounded by. We saw in the development of histone modifications how dependent we were on the work with mutations in Drosophila to detect the changes in gene expression which are mediated by changes in histone methylation carried out by the polycomb and trithorax complexes. Even more staggering are the observations which were made in Drosophila when mutations in a couple of genes changed the body structure of the fruit fly. In one case, the segment which should have carried antennae carried legs. It turned out there was a cluster of genes which were termed *Hox* genes which, in development, specify positional identity of the body segments. The mutations in *Hox* genes identified in Drosophila resulted in complete changes in body structure and importantly, these genes are conserved in evolution.

They are ancient genes and a *Hox* gene cluster is found in our friend the worm C elegans. However, and perhaps not surprisingly, we of Homo sapiens, in common with other mammals and birds, have four clusters of *HOX* genes each found on a different chromosome. In development of the embryo, the HOX proteins specify and ensure that the correct structures form in the correct places of the body. This is clearly a very important function indeed and mutations can result in disorders of limb formation. In a cluster, the transcription reads right through all the genes in that cluster and the *HOX* genes are expressed at the correct time in embryonic development. In the intensive investigation of this domain of the genome it was found that many lncRNAs are interspersed between the *HOX* genes and are also expressed along with the *HOX* genes. One of these lncRNAs, called *HOTAIR*, was found to be expressed from one *HOX* cluster but controlled the expression of genes in another *HOX* cluster which was found on a different chromosome. This is a unique function performed by *HOTAIR*, i.e. to be able to work at a distance to regulate gene expression. The sequence of the *HOTAIR* gene is not well conserved in the mouse making it difficult to confirm the function *in vivo* in mouse models.

Although *HOTAIR* was discovered in this way, work has intensified on studying its function since a dramatic increase in expression of *HOTAIR* is seen in many cancers, particularly in breast and lung cancer (see Chapter 4).

MORT: A lncRNA connected to cell mortality (see Figure 2.6)

Experience has shown that normal human cells can only be cultured in the laboratory for a limited number of cell divisions – they are mortal. In contrast “cell lines” can be cultured from many cancers and grown indefinitely – they are "immortal" and can even be frozen using the right conditions and recovered to begin growing again. These cell lines form a crucial source of human cells which can be used to study basic aspects of cell metabolism and genetics as well as identifying features of specific cancers.

Perhaps more than in any other tissue, normal breast epithelial cells, which line the ducts and lobules of the mammary tree, and which will produce milk after pregnancy, have been cultured in the laboratory. This is mainly because these cells can be grown from milk produced at lactation and also from tissue available from women (with their permission !!) who have breasts which are too large for comfort and choose to have them reduced in size by reduction mammoplasty (RM) surgery. The removal of tissue from inner organs is not so easy and is generally restricted to individuals who are having surgery for removal of abnormal tissue. So, for the last 40 years or so scientists have been culturing normal breast epithelial cells which are the cell type in the breast which can give rise to breast cancer and Professor Martha Stampfer in Berkeley, California has pioneered this field.

Using the normal breast epithelial cells cultured from reduction mammoplasty tissue taken from different women, Martha and her laboratory have been investigating whether they can convert the mortal cells to become immortal by regulating specific genes. They found that they can do this in two stages. First when the cells begin to multiply more slowly, they meet the first barrier (stasis) from where they can be kick-started by downregulating the expression of a protein which blocks the cell cycle (p16 affecting the RB tumour suppressor RB, see Chapter 4 for a description of the cell cycle). The post stasis cells then go through more cell divisions before meeting another barrier which can be passed by expressing a proto-oncogene known to be increased in many cancers (*cMYC*). The cells then become immortal and comparing the gene expression of the cells at different stages, they find that a dramatic change is seen in the immortal cells – they no longer express a lncRNA which they have called *MORT*[22] (see Figure 2.6).

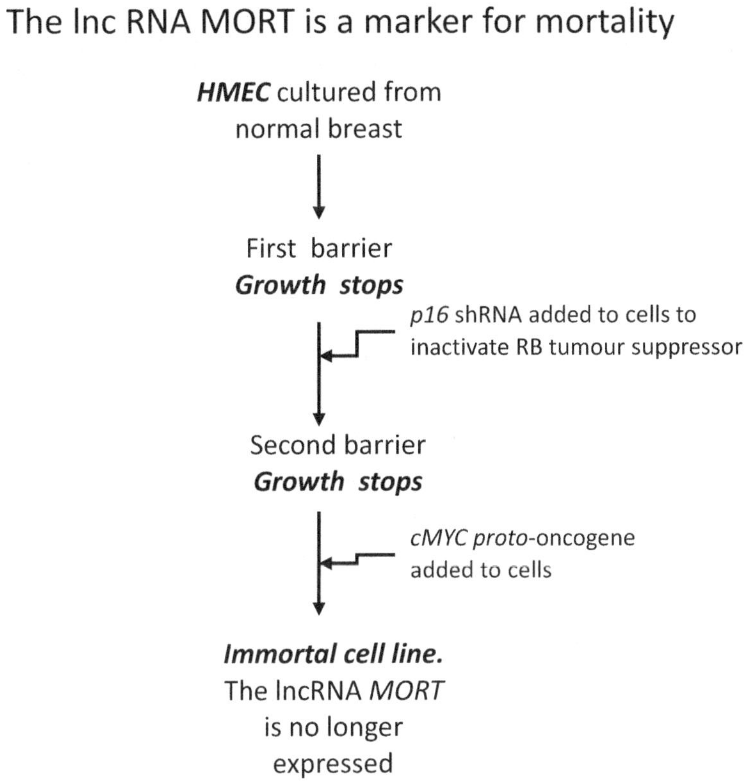

Figure 2.6: Human Mammary Epithelial Cells (HMEC) can be cultured from milk or from breast tissue but only for a limited period of time, they are mortal. *To ask whether these mortal cells could become immortal, breast tissue was cultured. When the cells stopped dividing, expression of the tumour suppressor p16 was silenced by adding a short RNA (shRNA) with a complementary sequence. At the next barrier to growth, the cMYC proto-oncogene was added. This combined treatment inactivated the tumour suppressor RB and increased expression of cMYC which resulted in the cells being able to be cultured in the lab indefinitely. The main difference in the expression of genes by these immortal cells was that they no longer expressed the lncRNA MORT*

The lncRNA *MORT*, named obviously for its presence in all mortal cells, is not expressed in the immortal cells because the CpG islands in the *MORT* promoter have been methylated thus down regulating gene expression. Referring to the publicly available DATA base, The Cancer Genome Atlas (TCGA), they find that *MORT* is not only silenced in breast cancers, but it is also reduced or silenced in the other common cancers with expression being inversely related to DNA methylation of its promoter.

Data from publicly available data bases also show that the *MORT* gene is only found and expressed in higher primates (chimpanzee, gorilla, and orangutan), and only incomplete sequences are found in Old World monkeys (baboon, rhesus). Therefore, *MORT* seems to be an evolutionarily young gene appearing late in primate evolution. In fact, up to one third of the lncRNA genes have been reported to be restricted to primates,[23] which means that the mouse models usually used to study the function of a gene *in vivo* are not available. Nevertheless, the emergence of MORT in primate evolution and its absence in multiple cancers suggests that it may be a tumour suppressor and useful to organisms like the primates who have a long-life span. The expansion of functional RNAs in higher primates provides a crucial dimension in the control of gene expression which could be required for the dynamic stability of their increasingly complex tissue development. Documenting the role of these RNAs in development, will be of great interest and the aberrant function of lnc functional RNAs in human disease is now an expanding area of research.

Non-coding RNAs bring Structural Order in the Nucleus for Control of Gene Expression

nDNA is packaged in the nucleus and the first stages of gene expression occur there to produce the different types of RNA. Some of these RNAs will stay in the nucleus and some, such as messenger RNA (mRNA) and microRNA (miRNA) will leave the nucleus to function in the cytoplasm to produce proteins (mRNA) or block their translation (miRNAs). Some of the proteins, such as those which are involved in control of gene expression, will return to the nucleus to exert their function. How are the multiple factors involved in controlling gene expression to be assembled? In Chapter 1 we described the nucleus as a "porous bag" and in one respect this can be justified as it is enclosed by a membrane – the nuclear membrane - and is surrounded by cytoplasm. Factors going out of or into the nucleus have sequences which are recognised by the nuclear membrane, which allow movement in and out. But what of the organisation within the nucleus? Do all the factors rush about hoping to bump into other factors needed for their function in controlling gene expression? Doesn't sound like a good plan, and very recent publications are now reporting on findings which are indicating that, in fact, the nucleus has a sub structure which is basically created by some ncRNAs[24].

With the recent advances in technology, scientists have been able to dissect interactions between the factors involved in transcription. With this 'SPRITE' technology, they have found that a large number of non-coding RNAs stay with the chromosome they are transcribed from and recruit other RNAs and proteins to form "territories", within which gene expression can be controlled. These are very recent discoveries and a full dissection of the function of these "territories" will take some time, but their identification is a landmark in understanding gene expression.

Changes in the Epigenome in human disease are now being documented, and these changes are related to the tissue of origin. Cancer however can develop in a large variety of tissues and is clearly a proliferative disease showing an imbalance in the relationship between proliferation and differentiation. Therefore, epigenetic changes are expected to play an important role in cancer and identifying these changes could lead to the development of novel therapies since changes in the epigenome are potentially reversible. In Chapter 4 we will examine some of the epigenetic changes which occur in cancer and ask if some of these changes are leading to clinical improvements in cancer management. Before that however we will, in the next chapter, return to our mitochondria and their DNA to discuss their function and the interactive epigenetic mechanisms which have developed which integrate the mitochondrial and nuclear DNAs.

Glossary for Chapter 2

CRISPR	This is a genetic engineering tool used by scientists to edit the DNA and can therefore modify, delete or correct precise sequences of DNA. It was originally discovered in bacteria where it is used as a defence mechanism against viruses.
Differentiation	Applies to a cell which is changing its phenotype epigenetically to take on the function of the tissue it comes from. After a while this differentiated cell will die and be replaced by another functioning cell. Differentiation is often blocked in cancer.
Embryology	The study of the embryo.
HOX genes	Code for transcription factors which ensure that the correct structures form in the correct places of the body.
In vitro	Studies carried out in the laboratory, outside of the organism. Latin for "in glass".
Long non-coding RNAs (lncRNA)	ncRNAs with a length of more than 200 nucleotides.
Messenger RNAs (mRNAs)	RNAs which code for proteins.
Methyl transferases	Methyl transferases are enzymes that can recognise a particular sequence of DNA or histones. In histones, they methylate one of the amino acids near that sequence which is usually a lysine and so they are termed lysine methyl transferases (KMTs).
microRNA (miRNAs)	A miRNA has a sequence that can block the function of a mRNA which has an antisense sequence.

Mutation	A change in the sequence of DNA either in a single base or by a deletion of insertion of bases. Some mutations affect the function of a gene, i.e. the protein coded for by the gene or the non-protein coding RNA expressed by the gene. Not all mutations affect function.
Non-coding RNAs (ncRNAs)	All of the RNAs which do not code for protein.
Palaeontology	The study of the ancient rocks and fossils of the earth. From the Greek: "Palaeo" means old and "ology" study of.
Phenotype	Defines the observable characteristics resulting from the controlled expression of genes. From the Greek "Phenomai"- how I appear to be.
Pluripotent cell	A cell that can divide to produce cells which can differentiate in multiple directions making different tissues.
Proliferation	Refers to the production of more cells. The word growth can include proliferation but may also refer to the expansion in cell size.
Stem cells	Stem cells present in each adult tissue. They divide slowly to give one new stem cell and a "progenitor" cell with an epigenetically different phenotype which then divides more rapidly before differentiating to be functional for a while.
Xist	Is a lncRNA, expressed from and involved in the silencing of genes on one of the X chromosomes in females.
Zygote	The cell which is formed by the fertilisation of an egg by a sperm. It is pluripotent, i.e. can form all the different cells which make up the new life.

Chapter 3
Mitochondria:
Power Houses of the Cell
with their Own DNA

"We are not made up, as we had always supposed, of successively enriched packets of our own parts. We are shared, rented, occupied. At the interior of our cells, driving them, providing the oxidative energy that sends us out for the improvement of each shining day, are the mitochondria, and in a strict sense they are not ours".

Lewis Thomas

We have, till now, talked about the DNA which is found in the nucleus (nDNA) on chromosomes, and this is the major constituent of our epigenome. There is however a set of organelles - the mitochondria - found in the cytoplasm of the cells of mammals (which of course includes us), that have their own genome, the mitochondrial DNA (mtDNA). Looking at all the other forms of life in existence today, most of these life forms also have mitochondria in their cells, and many, like mammals, have their own DNA, which brings with it a very effective method of producing energy using oxygen. These mitochondria have been called "the powerhouses of the cell" as we depend on them for burning the food we eat to generate the energy we need for carrying out all the cells' biological functions. There are many mitochondria in each of our cells, (except red blood cells) the number varying with different tissues, but all our cells have several and therefore multiple copies of their DNA. Despite this, mtDNA constitutes only 1% of the total DNA in the cell. Despite the smaller contribution in terms of numbers of genes, their role in epigenetics is very important throughout the body as they have learned to collaborate with the nuclear DNA and control the expression of some of the genes in nDNA.

Origin of Mitochondria in Evolution

As discussed in Chapter 1, there is general agreement that mitochondria developed from a bacterium which entered a different cell and stayed there as a permanent guest., The idea was first put forward by a botanist, Emanuel Wallin in the late 19th century. However, his work was largely ignored by the scientific community until 1967 when Lynn Margulis (then Lynn Sagan), published a landmark paper providing support for her "endosymbiotic" (where one organism lives within another for mutual benefit) theory, proposing that different cells could combine to produce a cell with infinitely improved biological potential. Lynn Margulis was refused acceptance for publishing her paper by fifteen journals before it was accepted by the Journal of Theoretical Biology[1]. Since then, she has been honoured as the "most successful synthetic thinker of modern Biology", although the journal Science has referred to her as "Science's unruly Earth Mother"! In 1983, she was elected to the prestigious National Academy of Sciences in the USA, and she was awarded the Darwin-Wallace Medal of the Linnean Society of London.

This endosymbiotic activity, which she proposed, is closely related to the development of a new kind of cell and takes us to a very general view of the forms of life on earth, and when this integration of their potential might have happened. In fact, there are two main forms of life on our planet, the *prokaryote and the eukaryote*[2]. *Prokaryotes* are single free-living cells, simple in structure without a nucleus to contain the bulk of their DNA. There are two classes of prokaryotic cells (bacteria and archaea), but it is bacteria which entered into a different host cell and stayed there as houseguests. The second type of life on earth are *eukaryotes* that have a membrane-bound nucleus containing their DNA (nDNA) which, as we have seen in Chapter 1, is bundled into chromosomes. The DNA of bacteria is simply a ring structure in the cytoplasm, and the mitochondrial DNA (mtDNA) found in our cells, is structured in this same way, supporting the concept that mitochondria developed from the bacterial house guests (see Figure 3.1).

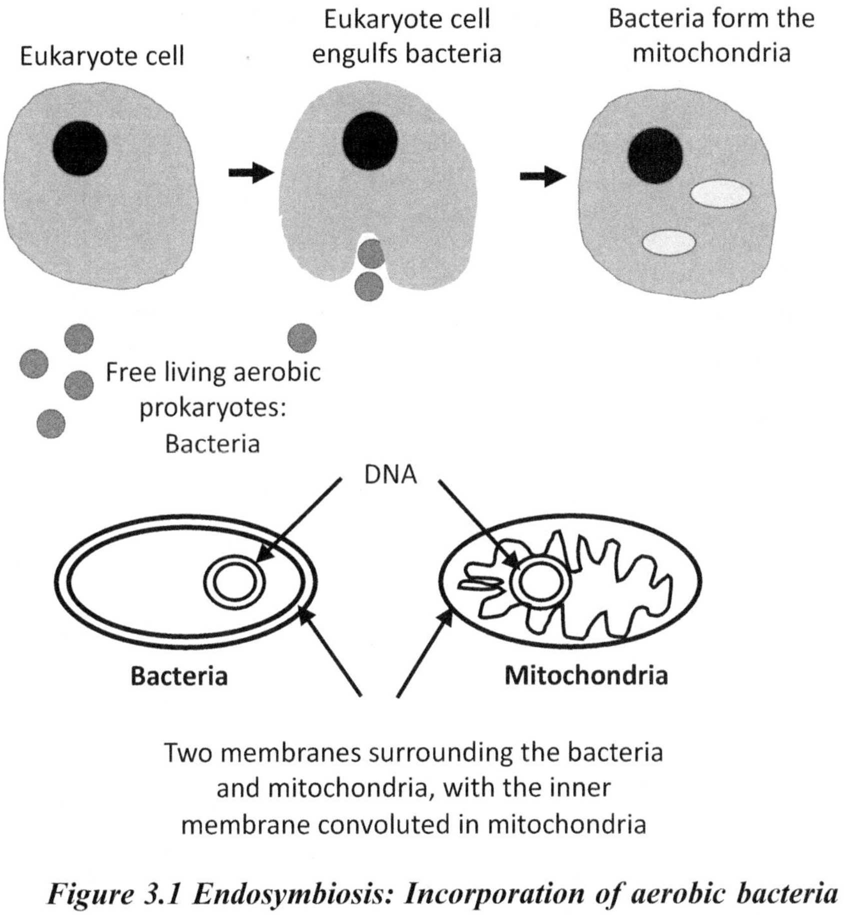

Figure 3.1 Endosymbiosis: Incorporation of aerobic bacteria into eukaryotes. *1.5 billion years ago, eukaryotes, which were developing at that time, allowed bacteria to enter their cells, bringing their own DNA. The bacteria stayed permanently and became the mitochondria, and the bacterial DNA became the*

mitochondrial DNA (mtDNA). The inner membrane of the mitochondria became convoluted to allow the surface area needed for oxidative phosphorylation (Fig. 3.2). Down the years the nuclear DNA in the nucleus (nDNA) and the mtDNA of the eukaryote cells became interdependent in an "endosymbiotic "relationship.

Most of the eukaryotes we know of today have mitochondria in their cells, and the ones present in our cells (and in all mammals and some other animal types), brought with them the genes coding for an amazing system for generating energy using oxygen, (known as oxidative phosphorylation). However, some other types of eukaryotes have mitochondria which have no DNA of their own and /or do not use oxygen to produce energy, i.e. they are anaerobic. Even though there are these differences in the function of mitochondria in different organisms, it is generally believed that endosymbiosis and the development of mitochondria in eukaryotes happened around 1.5 billion years ago, around the time eukaryotes appeared[3].

It is very interesting to note that at about this time, (estimated 1.5-2.0 billion years ago), another endosymbiotic event was initiated which resulted in the development of chloroplasts in some plant eukaryotes. In this event, bacteria (known as cyanobacteria, a class of bacteria that can photosynthesise) entered some plant eukaryotes to give rise to these organelles which can harness the sun's energy. Amazingly they carry out a reaction where CO_2 and water (H_2O) react to form oxygen (O_2) and glucose (food for the plant), a process known as photosynthesis. The ability of these chloroplasts to produce O_2, provided an awe-inspiring collaboration of the two endosymbiotic events where the chloroplast produces the oxygen for the mitochondria, to provide energy through oxidative phosphorylation (see below). An even more spectacular development in both examples of endosymbiosis is that with evolutionary time, most of the genes originally present in the incoming bacteria were transferred to the nuclear DNA (endosymbiotic gene transfer)[4]. Only a few genes of the original bacterial DNA stayed in the new organelles. This does not mean that all the proteins from the protein coding genes now in nDNA could not enter the chloroplast or mitochondria. On the contrary, in both cases the membranes of the organelles allow their import into the stroma and matrix of the chloroplast or mitochondria. Most organelles in eukaryotes are to a large extent controlled by the nucleus and the endosymbiotic gene transfer to the nucleus seen with mitochondria and chloroplasts has led to an integrated, eukaryotic genetic system under the regulatory dominance of the nucleus. Nonetheless the integrated system goes both ways and epigenetic modification of the transposed genes can be regulated by signals from the organelle (see below).

The development of endosymbiosis is argued by Margulis to have been a hugely important step in the development of multi-cellular eukaryotes, even though it happened so long ago. Indeed, the event is the most significant to have occurred in the development of Life. Whatever the original cells were that created the new eukaryotic cell with a nucleus and mitochondria, and or chloroplasts, they gave rise to the emergence of all the multicellular organisms we know of today, including ourselves[5].

When eukaryotic cells divide, the DNA in the chromosomes is carefully segregated to provide each daughter cell with identical copies of the nDNA (see Chapter1). The introduction of sex involving two parents to produce a new organism also brought in a new dimension by allowing recombination of their nDNA to occur in the germ cells, whereby pieces of DNA break and recombine in different position. This introduces variation which can further evolution. Bacteria however divide by fission to produce new cells and so do our mitochondria, which developed from them and mitochondria and their DNA are only inherited from one parent, our mothers, (see below).

At about the same time as Margulis published her paper, another scientist, Kwang Jeon, who was working with amoeba, found that endosymbiosis could be observed in his laboratory. Amoebae are single- cell organisms which can move and change shape by putting out extensions called pseudopodia (false feet). They are simple eukaryotes and have more than 20 chromosome pairs, but some amoebae have mitochondria, and some do not. Kwang Jeon's story is a bit like the accidental discovery of penicillin, where Alexander Fleming noticed that the plates of bacteria that he was growing died when they were close to a mould which had infected the plates. The mould of course was producing penicillin which killed the bacteria. In Jeon's case, his cultures of amoeba became infected with a bacterium and most of the amoeba died, but a few survived and grew up to make - after a few months - a new culture where the cells had intracellular bacteria which had entered the amoeba. He could show that the amoebae were now dependant on these bacteria, since when he gave them antibiotics the cells died although the original not-invaded cells did not[6]. Jeon's observation could be interpreted as a possible example of an early event in the pathway to the development of a eukaryotic cell with both a nucleus, chromosomes and mitochondria.

It is important to realise that in the remote time of 3.7 billion ago, when simple forms of life developed, to the time 1 to 2 billion years later when the eukaryotic cells appeared, with DNA-containing mitochondria (and chloroplasts in some plants), there were dramatic changes occurring in the levels of oxygen in the earth's atmosphere and in her seas. This variation had a dramatic effect on whether life depended on oxygen for respiration or not. In fact, oxygen could be toxic to the cells without the correct enzymes and pathways available to utilise the oxygen to produce energy.

Thus, early forms of life developing when oxygen was low, used "anaerobic" pathways (without oxygen) to produce energy, but as oxygen levels increased in the atmosphere and in the seas of earth, newer and much more effective ways of producing energy developed. It is at this stage that the road was opened to combine the eukaryote (with its nucleus and nDNA) with a bacterium which had DNA coding for the highly efficient system for producing energy using oxygen.

Powerhouses of the Cell: The Production of Energy

In mammals and in many complex multi-cellular organisms, the invading bacteria, which became mitochondria, brought with them the incredible system of "oxidative phosphorylation" to produce the energy that all the cells of the body need. Starting with glucose which comes from our food, particularly from starch and other glucose polymers which we can digest, there are several stages which lead to the production of energy which is finally stored in the molecule called ATP (adenosine triphosphate). ATP consists of the base adenine, a ribose sugar and three bound phosphate groups and carries the energy to where it is needed. Here the last phosphate is removed, releasing energy. The system is mind boggling in its complexity leading to the simplicity of how ATP releases energy. With oxidative phosphorylation the cell can produce 15 x more ATP than it could do under anaerobic condition, without oxygen and without mitochondrial genes (see Figure 3.2).

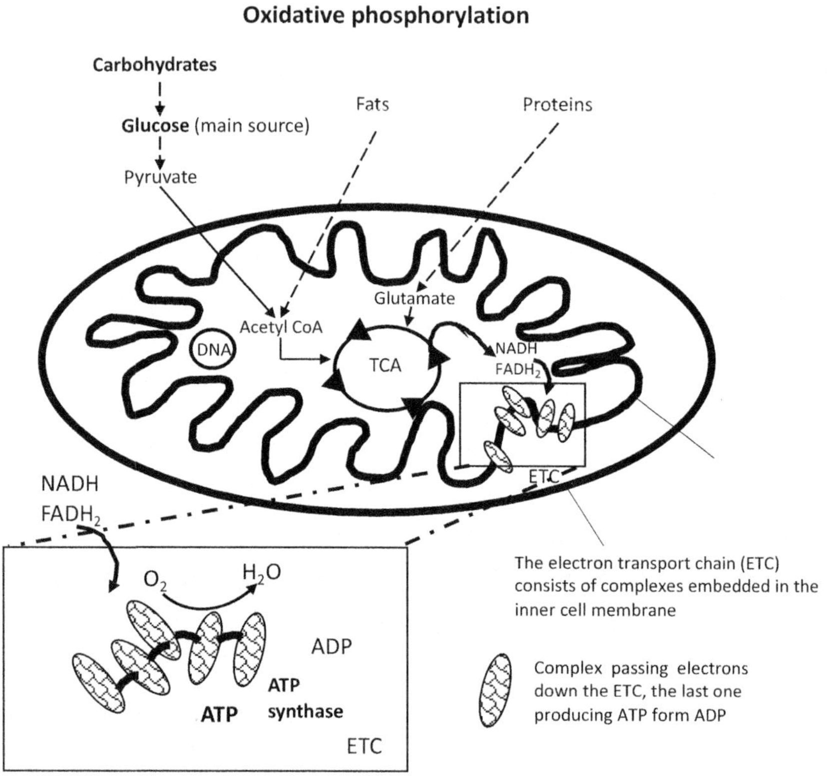

Figure 3.2: Oxidative phosphorylation takes place in the mitochondria to produce ATP. *Glucose, which is the main source of energy, is converted to pyruvate which in turn is converted to acetyl CoA in the mitochondria. Acetyl CoA is also produced from fats and enters a cycle, called the TCA cycle, consisting of chemical reactions which produce the important molecules NAD and $FADH_2$. Acetyl CoA is an important compound as it also serves as a substrate for the acetylation of histones within the nuclear chromatin. In the electron transport chain (ETC) NADH and $FADH_2$ act as a source of electrons. Oxygen is reduced, by the addition of electrons, to water and ATP is synthesized from ADP by the enzyme ATP synthase by the last complex in the ETC, ATP is the source of energy for all the cells in our body and 32 molecules of ATP can be*

produced from one molecule of glucose in the mitochondria using oxidative phosphorylation.

The number of mitochondria in a cell is different in different tissues and in the different stages of differentiation of the tissues, expanding from a low number in the stem cells to higher numbers as the cells differentiate to become functional (see Figure 2. 2 in Chapter 2 for cell differentiation). Where the tissues require high energy production, as for example in muscle, brain and heart, there are many more mitochondria in these cells compared to other cell types. For example, heart muscle can contain around 5,000 mitochondria whereas as spleen cells contain about 100 and sperm about 50-75 mitochondria. Moreover, where the demand for energy production is high, the mitochondria position themselves intimately within the structure of the tissue. Figure 3.3 illustrates this for muscle and for a sperm cell which is in the race for fertilising an egg, and the tail needs the energy to get there. Incidentally, muscle training/conditioning, for example push ups and squat thrust exercises, increases oxidative phosphorylation, increasing the number of mitochondria resulting in more ATP being produced in skeletal muscles.

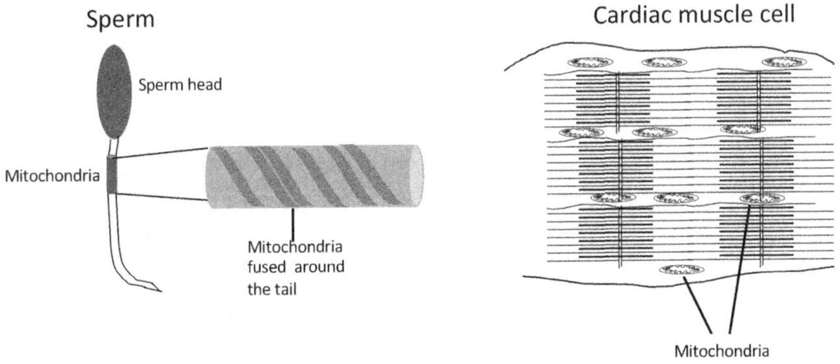

Figure 3.3: Mitochondria can adapt to the tissue: *The sperm requires energy produced by the mitochondria for its race to fertilize the egg. The mitochondria wrap around the tail of the sperm that*

after fertilization falls off, so the mitochondria are lost, so only the mother's mitochondria are in the fertilised egg. Muscle cells require a lot of energy and so contain many mitochondria. Exercise can increase the number of mitochondria within the muscle cells.

Mitochondria can make the Proteins Coded for by their own DNA inside the Organelle

Mitochondrial DNA is a closed circular double stranded structure with most of the genes expressed being transcribed from one strand. The DNA is not wrapped around histones and therefore does not contain nucleosomes as the nuclear DNA does. mtDNA codes for only 13 proteins which are components of the complexes which carry out their unique function of oxidative phosphorylation. They do however also have genes coding for some non-protein coding RNAs, ie 22 transfer RNAs and 2 ribosomal RNAs[7]. With these RNAs, they can, within the mitochondria, translate the 13 protein coding genes to make the relevant proteins, most of which are involved in energy production. However, the mitochondria do need help from the nDNA which has genes coding for other proteins necessary for the mitochondrial energy - producing function as well as for their replication. In fact, there are around a thousand proteins required by the mitochondria which are produced by the nDNA, and many of the genes coding for these protein were originally in the mitochondrial DNA but with time they were transferred to the nuclear chromosomes[8]. The nuclear gene coding for the polymerase POLRMT is responsible for the transcription and replication of the mtDNA and also helps with translation of the mtDNA coded proteins[9].

The proteins produced by the nuclear DNA are imported by the mitochondria, but the level of their expression is determined by the nuclear chromatin and is affected by changes in epigenetic modifications including nDNA methylation and the chemical modification of their associated histones. As we will see below, these epigenetic changes in nDNA, modifying gene expression are governed by the mitochondria. Thus, there is a two-way interaction controlling nuclear and mitochondrial DNA function, see Figure 3.4.

Are Mitochondrial Genes under Epigenetic Control?

As we have seen, mitochondrial DNA is not wrapped around histones as is nDNA, and for many years it was thought that mtDNA was not methylated and so had no means of epigenetic control of the mitochondrial genes expressed. Indeed, in the 1970s, papers were published reporting that mtDNA was not methylated and the debate has continued for many years. However, as techniques have improved to detect DNA methylation it has become clear that there is methylation of some cytosine bases in mtDNA although at a lower level than in nDNA. However, contrary to the early reports of methylated mtDNA, it is now believed that this methylation occurs more often on cytosines that are not in CpG islandsthis methylation of mitochondrial DNA achieved as there are no enzymes able to methylate mtDNA produced by mitochondria DNA? What appears to be happening is that enzymes that are involved in the addition of methyl groups to nDNA and are coded for by nDNA contain a special signal in their protein sequence called a mitochondrial targeting sequence (MTS) that can additionally direct them to the mitochondria. So, when these enzymes are delivered to the mitochondria, they can methylate cytosines within mtDNA.

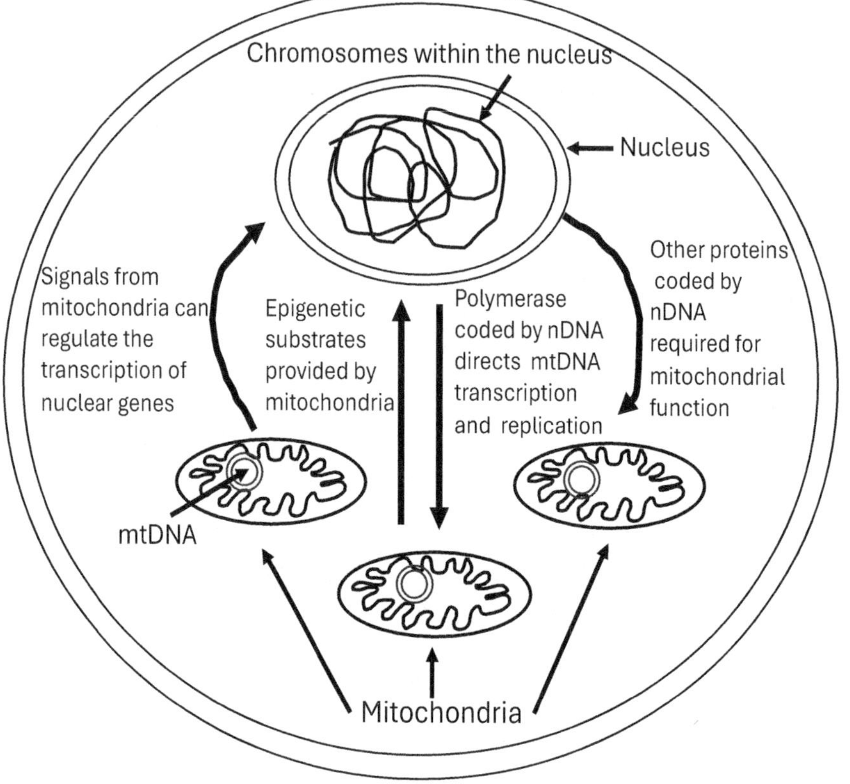

Figure 3.4: Mitochondria and nDNA talk to each other. *Proteins encoded by nDNA are required for mitochondrial function, like the polymerase POLRMT which is crucial for transcription and replication of mtDNA (see text). On the other hand, besides supplying ATP from oxidative phosphorylation and substrates for epigenetic regulation, like acetyl CoA and SAM, signals from the mitochondria can affect transcription of the nuclear genes. See section on Dynamic Interactions between the nucleus and the mitochondria for definitions and roles of acetyl CoA and SAM.*

The question is, does methylation of mitochondrial DNA affect transcription of its genes to give the RNAs and proteins which are essential for mitochondrial function? The studies aiming to answer this question have not been going on for as long as studies analysing the effects of methylation of nDNA on gene expression. Moreover, the mitochondrial studies are made more complex by the fact the polymerase and factors working with it for transcription of mitochondrial DNA are in fact nuclear genes – as are the DNA methyl transferases. The gene coding for the polymerase (POLRMT) acting on the mtDNA is found in the nDNA and is very different from the polymerases (also coded by nDNA) which function in controlling nuclear DNA expression (see Chapter 1). In fact, POLRMT resembles a polymerase found in a virus which infects bacteria!! It is likely that the gene coding for this polymerase and the nuclear genes working with it, originally came into the eukaryotic cell with the bacteria and have been transferred to the nuclear genome (see Figure 3.4). We will see in the next chapter, where we discuss the epigenetic factors in cancer, that POLRMT is now seen as a potential therapeutic target in some cancers.

While there are suggestions that methylation of one of the two strands of mtDNA may affect gene expression, this is still controversial. It is the methylation of the nDNA genes which affect the processes of mtDNA transcription, replication and translation which are defined and have profound effects om mitochondrial function.

Can Mitochondrial DNA Methylation serve another Function?

The fusion of the sperm and the egg is the first step in making an embryo which has the potential to form a new human being (see Chapter 5 for more information). The single cell, formed by the fusion of the egg with the sperm, divides a number of times to form a blastocyst and it is this blastocyst that implants itself into the lining of the womb and forms the embryo. The stage from the blastocyst to the implanted embryo requires a considerable amount of energy and so there is an increase in oxidative phosphorylation which is essential for the survival of the embryo once it is implanted in the uterus. However, this increased oxidative phosphorylation leads to an increase in reactive oxygen species (ROS) which are highly reactive and have the potential to damage both mtDNA and nDNA. Until recently it was unknown how the mtDNA was protected from this stress but now it appears that it maybe by methylation. Around the time of implantation, mtDNA undergoes extensive methylation that protects the mtDNA from damage caused by the reactive oxygen species[11].

Methylation of nuclear DNA in the embryo is controlled by two enzymes known as DNMT3A and DNMTB. Importantly, during the time around implantation these enzymes are transported into the mitochondria and are responsible for the methylation of mtDNA. This protects the mtDNA from the damage that can be induced by oxidative stress, through activation of ROS. Interestingly, again, the methylation did not occur in regions of CpG islands but was found within all regions of the mtDNA. Thus, in mitochondrial epigenetics, DNA methylation, plays a role in protecting the mtDNA from oxidative stress. Whether methylation of mtDNA influences expression of mtDNA genes is still not clear.

Mitochondrial DNA is Inherited from your Mother- and her Mother and her Mother...

It may come as a surprise to some readers to learn that, unlike nuclear DNA, which is derived from a male and a female parent, mitochondrial DNA is different: whether you are male or female, mtDNA is only inherited from your mother. This is because, when the sperm rushes to win the race to fertilise the egg, the mitochondria required for energy production are wrapped around its tail which is lost on fertilisation (see Figure 3.3). Therefore, all of the mitochondria in the fertilised egg (the zygote) which will give rise to all the cells in the new life, are from the mother. The maternal heritage allows the ancestry of any individual to be tracked through their mother's lineage.

Although the whole mitochondria with their DNA are not passed from the male parent to the embryo, recent data has found that small fragments of the transfer RNAs coded by mitochondria can be transferred. It has been known for some time that transfer RNAs can be cut into small pieces, not randomly but by specific enzymes to yield what have been called tsRNAs. These tsRNAs have been shown to have effects on cell function and cell proliferation. It is also known that the mitochondrial tsRNAs can be transferred from the sperm upon fertilisation to the egg and consequently are evident in the embryo. These data are very new and to study the effects of these mitochondria derived tsRNAs on the function of the embryo, single mice embryos have to be analysed, confining the study to highly technical research activity[12]. What is known is that these mitochondrial tsRNAs are susceptible to the environment. An observation which will be discussed in Chapter 5 is that when the male parent, has been on a high fat diet for a couple of weeks before conception, the function of the mt tsRNAs is disturbed.

While most people are interested in tracing their more recent ancestors using nuclear DNA, tracking with mitochondrial DNA delves into prehistory and you can find out where your struggling female ancestor was maybe 25,000 years ago. It is now possible for anyone to have their mtDNA sequenced and some readers might have arranged for this to be done by a company and already have the important sequence of their mtDNA. A considerable amount of data has now been accumulated on mtDNA sequences and it is possible to separate these sequences into groups on the basis of their mtDNA sequence. Indeed, as described by Bryan Sykes, the people of

Europe have been separated into seven groups, (the seven daughters of Eve) arising from Mitochondrial Eve who left Africa around 100,000 years ago[13]. Such groups are defined by specific sets of shared genetic variation which has resulted from benign mutations occurring down the years. These groups are termed the maternal haplogroups. It has become clear that the different haplogroups affect differently the level of expression of the nuclear genes which are required for mitochondrial function and replication. The surprising fact is that the difference in the mitochondrial haplotypes is reflected in the different patterns of methylation of the mitochondria-relevant genes in nDNA, and in the epigenetic modifications of their associated histones[14].

It should be said that most of the mutations which occur in mtDNA - as it is transferred from mother to daughter down the years, are mostly not pathogenic, but some can be, and there is a difference between haplogroups in their development of disease in general and in heritable mitochondrial disease in particular.

Identification of the Executed Russian Tsar and his Family: The Importance of mtDNA

Near the start of the Russian revolution, Tsar Nicolas II was forced to abdicate on March 15th, 1917. The royal family were held prisoners and in July 1918 they were herded into a cellar and the Tsar, Tsarina and their children were shot by the Bolshevik firing squad. Some were stabbed to death if the bullets had failed to kill them.

In 1991 nine skeletons were found in a shallow grave in Ekaterinburg, Russia. These skeletons, which contained bullet wounds and evidence of stabbing, were believed to be that of the Tsar, Tsarina, three of their five children, three servants and the Royal Physician, but evidence was required to prove this was the case. In 1992 the Chief Forensic Medical Officer of Russia, Pavel Ivanov, asked Dr Peter Gill if he would be interested in working with him on identifying the remains. Dr Gill who worked at the UK Forensic Science Centre, had previously worked with Alec Jeffreys on DNA profiling (see Chapter 1). Thus began the collaboration between British forensic scientists and a Russian forensic scientist to identify the remains found in a shallow grave[15].

Although very small amounts of nDNA could be recovered from the bones, using the technique pioneered by Alec Jeffreys of nDNA fingerprinting, five of the remains were shown to be that of a Father and Mother and three of their daughters. However, this did not identify the remains of being that of the Tsar and his family. To do this, DNA had to be compared to a living relative. In contrast to nDNA, mtDNA is more likely to survive for extended periods and indeed mtDNA has been analysed from skeletons 5,500 years old[16]. Moreover, as discussed above, mtDNA can be used to trace maternal inheritance and is the best and only option to test for connections between individuals that are generations apart. Thus, parts of the mtDNA were sequenced from all the individuals and compared to living relatives. One such relative was HRH Prince Philip, the Duke of Edinburgh, who was the great nephew of the Tsarina, and they both inherited mtDNA from Queen Victoria via her daughter, Princess Victoria (see Figure 3.5).

Prince Philip provided a blood sample and comparison of his mtDNA with that of the mother and the three children confirmed these to be the remains of the Tsarina and her three daughters as their mtDNA was identical with that of Prince Phillip's. The Father's mtDNA was compared to the mtDNA of two of his living relatives and showed the father to be the Tsar. Thus, nDNA and mtDNA sequencing solved the mystery: the bones found at Ekaterinburg were the bodies of the Tsar Nicolas II, the Tsarina and three of their daughters. The fate of the other two children remained a mystery until in 2007 another grave was found, close to one containing the Tsar and Tsarina. This grave contained two bodies which by mtDNA testing was shown to be that of the other two children who had inherited the same mtDNA as Prince Phillip.

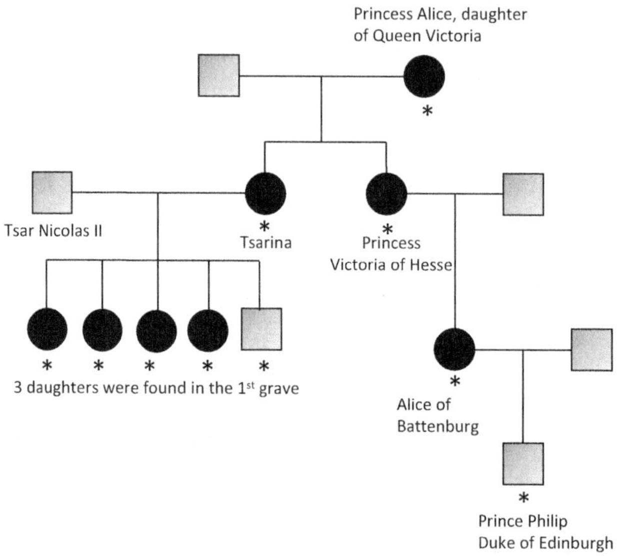

***Figure 3.5 Lineages of Prince Philip and the last Tzarina of Russia and her children**. Black circles represent females and grey squares represent males. Since mitochondrial DNA (mtDNA) is only inherited through the female lineage, the Tzarina, her children and Prince Phillip all inherited their mtDNA from Queen Victoria through her daughter, Princess Alice. MtDNA was extracted from a blood sample taken from Prince Phillip and from samples taken from the remains of bones buried in Ekaterinburg, Russia where the Russian Royal family had been killed many years earlier. The analysis showed that mtDNA taken from four bodies was identical with that of Prince Phillip (* indicates individual with identical mtDNA). These four were thus identified as the Czarina and 3 of her children.*

The Nucleus and the Mitochondria can talk to each other

It is now clear that nuclear DNA methylation has a strong impact on mitochondria for their content and function. For example, expression of a protein made by a nuclear DNA gene essential for mtDNA transcription, replication and translation, is epigenetically regulated by its methylation in the nuclear DNA. Thus, the gene coding for this protein is heavily methylated in stem cells that have few mitochondria, and their number expands as the cells differentiate as the methylation of the gene in nDNA is decreased in the functional cells, allowing for more transcription (see Chapter 2, Figure 2.2 for differentiation).

So, are there signals coming from the mitochondria to epigenetically modify the nuclear DNA and its associated histones? The fact that the DNA methylation patterns are different for different mtDNA haplotypes, as mentioned above, suggests that there are. One surprising route of interaction is the provision, by the mitochondria, of substrates used by the enzymes which epigenetically modify nDNA and their associated histones. One important component is acetyl CoA, (produced for oxidative phosphorylation) which is required for acetylation of histones which we know results in an open structure of nDNA, allowing gene expression. Also, the molecule which provides the substrate for methylation of DNA or of histones - SAM (S-Adenosyl methionine) is also provided by the mitochondria[17] (see Figure 3.4).

Mitochondrial Dynamics: Fission, Fusion, Mitophagy and Mitochondrial Transfer

The continuous production of new mitochondria for growth and differentiation means that the mtDNA has to be replicated many times, and the rate of mutation in mitochondrial DNA is 100x that seen in nuclear DNA. Fortunately, they have a repair system, which can help to reduce the expression of these mutations. This system is similar to the one used by bacteria, supporting the hypothesis that the mitochondria developed from invading bacteria.

Apart from any changes in the sequence of mtDNA which might happen, the mitochondria undergo a continuous process of "fission, fusion and mitophagy" (division, fusion and destruction) which affects their shape, contents and number. These processes contribute to maintaining healthy functional mitochondria by removing damaged components or whole mitochondria with aberrant function. The mitochondrial organelles are enclosed by a double membrane, the outer membrane (OMM) separating them from the cytoplasm and the inner membrane (IMM), which is the site of energy production by oxidative phosphorylation, see Figure 3.2. These membranes allow the possibility of the fusion of two mitochondria which can then mix their contents, or they can allow fission of a single mitochondrion to produce two organelles.

The production of new organelles by fission is very important for expanding the number of mitochondria in normal differentiation and is crucial for the development of the embryo. The use of this process of fission to multiply themselves is another feature which links them to their origin as bacteria which also replicate by fission (Figure 3.6). While fission is necessary for the normal function of the mitochondria, it can be usurped if there is a mutation in the mtDNA that will then also be expanded, interfering with normal function. On the other hand, fusion of two mitochondria, one with mutated mtDNA, the other with normal DNA, allows mixing of their contents and the dilution of the mutated mtDNA. This contributes to maintaining mitochondrial homogeneity and to minimising the level of a pathogenic mutant DNA. There is another overall control mechanism – mitophagy- where autophagosomes (eating one's own body) selectively eat up and remove dysfunctional mitochondria.

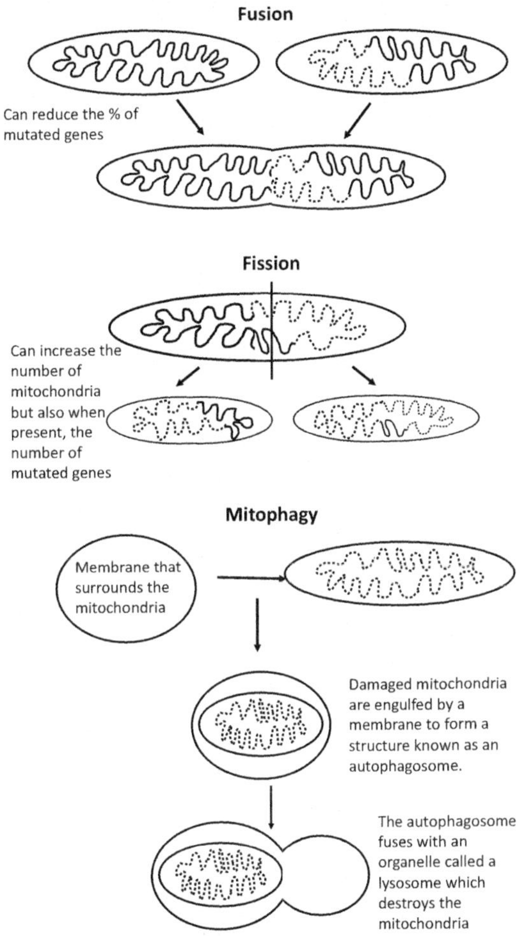

Figure 3.6: Mitochondria undergo fission, fusion and mitophagy. *Fusion, mitochondria can fuse to reduce their number and to dilute mutated genes. Fission, when a cell divides the mitochondria undergo fission to allow the same number of mitochondria in each daughter cell. Fission can also result in increasing the number of mitochondria with mutated genes but in*

contrast can segregate mitochondria with mutated genes allowing for these mitochondria to be destroyed by mitophagy.

Mitochondria also have another trick up their sleeve, they have the ability to move from one cell to another[18]. They can do this via a number of mechanisms but the most common is the forming of intercellular connections known as nanotubes through which mitochondria can be transported. Why should this occur? It is thought that if a cell suddenly requires increased energy, then the cell can import mitochondria from neighbouring cells. This has been referred to as the "generator hypothesis". Houses rely on the national grid for their electrical supply but if there is a fault in the supply, some households can use their own generators. Cells can rely on their "generators", which are mitochondria coming from other cells, until their own mitochondria are restored. Importantly cancer cells are known to import mitochondria from cells within the tumour microenvironment (see Chapter 4). It is speculated that this may affect the growth of the cancer cells and contribute to their evasion of the immune response, but further work is required to confirm and investigate the pathways involved.

Mitochondrial Diseases

It is perhaps not surprising that the bacteria brought with them, not only their incredible oxidative system for producing energy, but also a new set of diseases. With their heritage from bacteria, the mitochondria bring the possibility of stimulating the inflammatory immune response. It is well known that after trauma or injury to the body, a *systemic inflammatory response syndrome (*SIRS) appears. It has now been found that the mitochondria are the culprits for stimulating this response as they and their contents are fragmented and liberated from the cell by the trauma. The products expressed by mitochondrial DNA and other specific molecules found only in bacteria and mitochondria can then recognise receptors on some immune cells. These receptors are called *pattern recognition receptors* (PRR) which are there to recognise patterns in bacteria *(pattern-associated molecular patterns* or PAMPS), but also recognise mitochondrial products, which then stimulate the receptors to mobilise an inflammatory response[19].

There are also mitochondrial inheritable diseases which are caused by mutations in genes which affect mitochondrial structure and function, and these genes can be found in the mtDNA or in the nDNA. These are called primary mitochondrial diseases (PMDs) which are difficult to diagnose as they can affect several organs and tissues because of the mitochondria's crucial role in producing energy. Moreover, the severity of the symptoms depends on the extent to which the mutant gene dominates compared to the normal gene. Mitochondrial diseases that are due to mutations in mitochondrial DNA are of course only inherited through the female line see Figure 3.7. This can be controlled to some degree by the balance between mitophagy, fission and fusion. However, while sometimes the interaction of these activities can result in maintaining a low level of a pathogenic mutated DNA, or eliminating it, more often than not a condition emerges where a certain percentage of the mutated DNA is present, together with the normal mtDNA. Importantly, it is the differences in the level of this "heteroplasmy" (the amount of mutated mtDNA compared to normal mtDNA), occurring between patients with the same mutation, which affects the progression of mitochondrial disease and makes diseases associated with mutated mtDNA highly variable in phenotype and therefore difficult to treat. Estimates of the number of individuals with a genetic mitochondrial disease range from 1 in 200 to 1 in 4000, the variability reflecting how difficult it is to identify whether patient's symptoms are caused by impaired mitochondria. Nevertheless, it is now thought that mitochondrial disease is one of the commonest forms of genetically inherited disease, which can affect multiple tissues and National and International societies and Institutions are trying to tackle the problem of how to diagnose and treat these diseases[20]. Of course, mutations can also occur somatically, that is after birth (soma-Greek for body) but the genetically inherited ones have been more studied, particularly where the mutation is in a gene in the mtDNA.

One of the commonest mitochondrial diseases is MELAS (*mitochondrial encephalomyopathy, lactic acidosis and stroke-like episodes*), which primarily affects the nervous system and muscles. The length of the title for MELAS reflects the widespread nature of the symptoms which are neurological and life expectancy for these patients can be very low depending on the heteroplasmy. It is speculated that Chares Darwin (of “The Origin of Species” fame), had MELAS, as he was always ill with multiple symptoms, but he did live to be 73 years old, so if he did have MELAS, the heteroplasmy must have been low. He did worry that he might be passing on a genetic problem but if it was a mitochondrial disease, he needn’t have, because as a male parent he could not have passed it on to his children (see above).

We saw in chapter 1 that when proteins are synthesised by the ribosomes found in the cell cytoplasm, different transfer RNAs (tRNAs) are responsible for bringing each amino acid in the right order to be assembled to form a protein. In the same way tRNAs made in the mitochondria are crucial for the synthesis within the mitochondria of the proteins coded for by mtDNA. It has been found that 80% of the mutations associated with MELAS are found in a mitochondrial gene which codes for a particular mitochondrial transfer RNA termed $tRNA^{LeuUUR}$. This tRNA takes the amino acid leucine to the mitochondrial ribosomes for protein synthesis and the mutation results in the aberrant function of mitochondria and of nuclear gene expression.

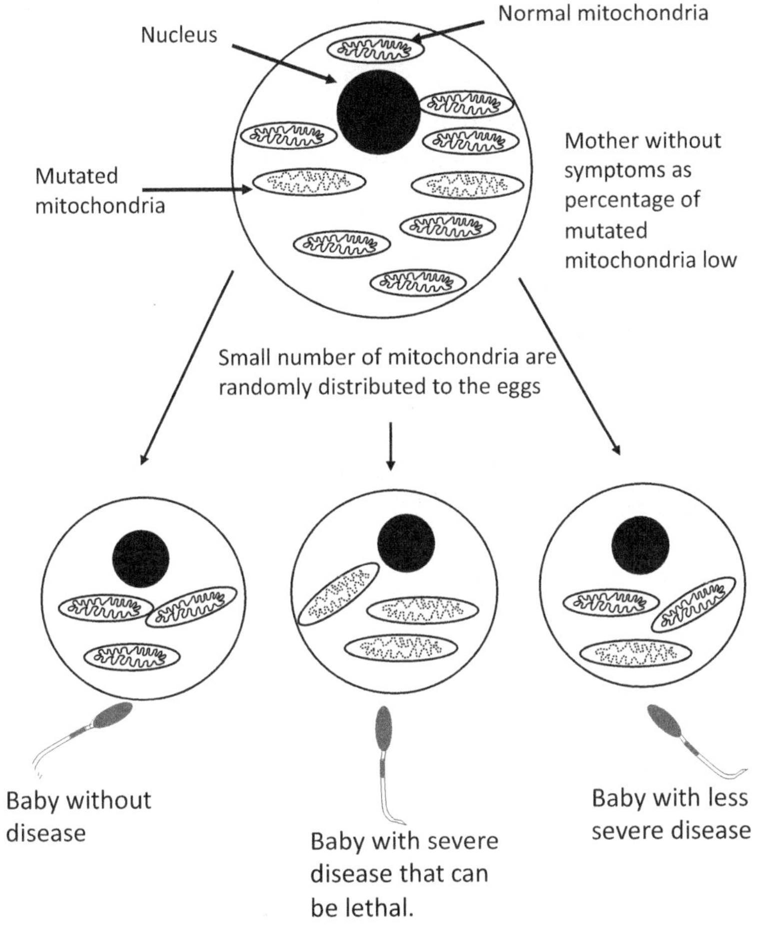

Figure 3.7: Mutated mitochondria DNA is inherited from the mother. *Some mitochondrial diseases are due to mutations in mitochondrial DNA. When the mother's eggs are formed, mitochondria are selected at random to enter each egg. The percentage of mutated mtDNA within each egg will determine the severity of the disease of the baby.*

The relative number of mutated tRNA to wild type (the normal gene), that is, the "heteroplasmy", determines the severity and symptoms of a mitochondrial disease. The data relating to MELAS show that with 20-30% of the mutant mitochondrial gene present, the patients usually develop type 1 or type 2 diabetes or show signs of autism. However, with 50-80% heteroplasmy of the mutant DNA, patients develop myopathy, cardiomyopathy or stroke-like symptoms either separately or together. These symptoms almost certainly relate to the fact that the tissues heart, muscle and brain require high energy production and are therefore very susceptible to aberrantly functioning mitochondria. With 90-100 % of the mutant present, perinatal lethal diseases are seen. There is no effective treatment for MELAS and the effects on the mitochondria of this mutation has been worked on in preclinical studies using cell lines, hoping to learn more about the mechanisms underlying the differences in disease development[20,21].

A mutation in a mitochondrial tDNA affects nuclear gene expression through inhibiting the supply of substrates for their epigenetic modifications

Scientists have been able to use cell lines with the same sequence of nuclear DNA, but with different heteroplasmy of the MELAS $tRNA^{LeuUUR}$ mutation to investigate the mechanisms underlying the different disease progressions seen in patients. They find that patients with different levels of heteroplasmy for the mutated mtDNA show different patterns of nuclear DNA gene expression, indicating epigenetic regulation[22]. Using this system, they have been trying to identify signals coming from the mitochondria which epigenetically regulate the expression of nuclear genes[22]. As stated above, the mitochondria are responsible for providing the nuclear chromatin with substrates for the enzymes which epigenetically methylate the nuclear DNA and add acetyl or methyl groups to associated histones. The scientists therefore looked at the level of the substrate acetyl CoA, required for adding acetyl groups to histones, and found that the levels of acetyl CoA given to the nucleus decreased as the levels of the mutant DNA found in the mitochondria increased. This was reflected in decreased acetylation of the nuclear DNA. The most dramatic result was seen when the level of mutated mtDNA was 90-100% (which can be lethal in children). In this case, the supply of acetyl CoA for histone acetylation was strongly decreased thus decreasing gene expression. When the level of expression of the mutant was lower (20-30%) where patients develop diabetes or show signs of autism, expression of both the mitochondrial and nuclear genes required for energy production was decreased. This effect correlates well with the serious mitochondrial energy deficiency seen in type 2 diabetes and in autism.

It has been difficult to develop therapies for mitochondrial disease and existing treatments has focused on supportive and preventive approaches, with few targeted therapies being available. Currently attempts are made to increase the energy producing function of the mitochondria by encouraging exercise to stimulate an increase in their numbers and function. Also, various food supplements are recommended. However, there have been few attempts to apply pharmacologically based therapy in the clinic Nevertheless, since the year 2000 up to 2020, there has been an exponential increase in the number of clinical studies registered for evaluating treatment, so that now, there over 200 studies on going.

Because inherited primary mitochondrial disease can be so severe, preventing transmission of the mutated gene from parent to prospective children is considered to be highly desirable. Without intervention, parents are faced with the choice of deciding not to have children in case one of them carries the mutant gene or going for adoption. Oocyte donation from an uninfected woman, providing healthy mitochondria is probably the least complex form of intervention, if this is acceptable to the individuals involved. It requires only the *in vitro* fertilisation by the father of an egg from a doner with healthy mitochondria. However, the prospective mother has no genetic input into the embryo and the new life, but with the advance of technology, other possibilities are becoming possible.

Prenatal testing

Compared to prenatal testing for Mendelian genetic disorders coming from mutations in nuclear DNA, prenatal testing for mtDNA defects is faced with a problem. This problem arises from the fact that differences in levels of the mutant gene give rise to differences in the severity of disease, which makes it difficult to predict the severity of the disease for each individual embryo carrying the mutant gene without knowing the level of heteroplasmy (see above).

"Three parent babies"

It has now become possible to carry out Mitochondrial Replacement Therapy (MRT) where an egg containing the nucleus of the biological mother but with healthy mitochondria from a female donor is fertilised *in vitro* by sperm from the father. To achieve this, two approaches have been taken. In the first, oocytes from both the mother and from the mitochondria-healthy donor are fertilised by sperm from the father. Then the nucleus from the donor oocyte is removed and the nucleus from the egg of the mother is added to the donor oocyte with healthy mitochondria. Alternatively, the nucleus from the donor cell is first removed and replaced by the nucleus of the mother before being fertilized by the father's sperm, (see Figure 3.8). It is worth mentioning that this technique of moving nuclei and chromosomes around was first developed by

John Gurdon many years ago working with lower organisms which make larger eggs[23] and it is with increases in microscopy technology that it has become possible to work with human eggs. Sir John must feel rewarded for what his hard work has led to, namely a technology for allowing parents, who had lost children due to inheritance of a mutated mitochondrial gene, to have a child with heathy mitochondria.

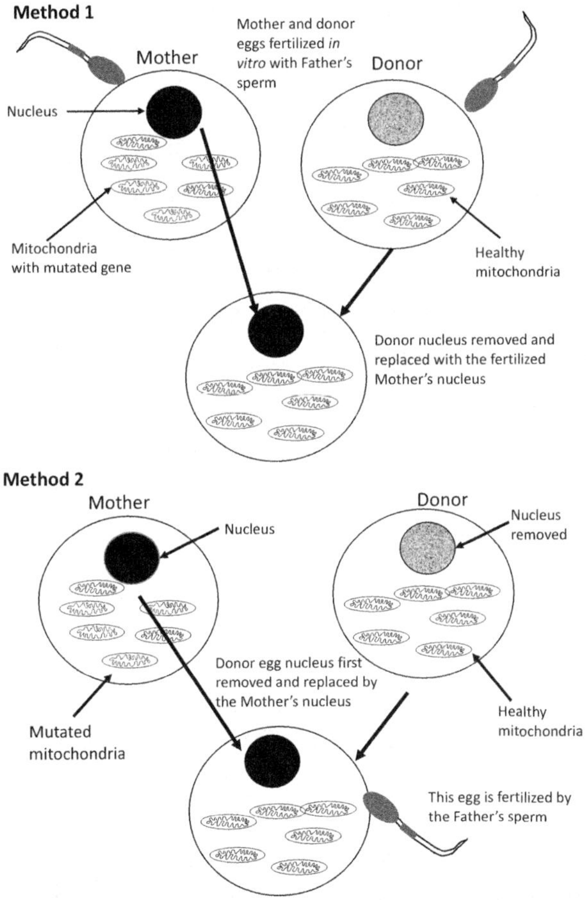

Figure 3.8: Methods for accomplishing 'Mitochondrial Replacement Transfer' (MRT). *In the first method, eggs from the prospective mother carrying mutated mtDNA and eggs from a donor carrying healthy mitochondria are fertilised by sperm from the*

prospective Father before replacing the nucleus in the donor egg with the nucleus of the prospective Mother. In the second method, the nucleus of the donor egg is removed and replaced by the nucleus from the prospective Mother before fertilization by the Father's sperm. With both methods, the egg which will give rise to the new life has been fertilized by the Father's sperm with its nDNA, has a nucleus from the prospective Mother with her nDNA, and has healthy mitochondrial DNA from the donor.

Understandably there are different opinions about the ethics of manipulating the contents of human germ cells, as any changes will be inherited by future generations. So, it is appropriate that we outline how long it took for getting to the point of passing a law making the clinical use of MRT allowable. The necessary legalisation for performing clinical trials for MRT was first put in place in the UK in 2015 following 17 years of discussions with experts and with the public, and when the techniques for making MRT possible were being developed.

In 1990, the Human Fertilisation and Embryology Act was passed in the UK. With commendable foresight, the Act allowed the inclusion of possible future developments in embryo research to be considered, and the Human Fertilisation and Embryology Authority (HFEA) was established. The Wellcome Centre for Mitochondrial Research in Newcastle University played an important role in furthering the pathway to the clinical use of MRT and further consultations with experts and the public led to a report from the Nuffield Council of BioEthics supporting the clinical use of MRT, but it was only in 2013, after more reports were collected and public views were again sought, that HFEA advised the government that there was "general support for permitting mitochondrial donation" in the UK. At this point the Government published draft regulations to amend the 1990 Act to permit the "clinical application of mitochondrial donation: These came into UK law in February 2015. Clearly the forces driving the legalisation of MRT were running parallel to the technological developments in working with lower organisms and human germ cells[24] and in 2016 license was granted to the Newcastle Fertility Centre at the Newcastle on Tyne Hospital NHS Foundation Trust. They are now following closely babies who were born using the MTR technique. Australia also has had the legislation in place approving the use of MRT since December 2021 and some countries, where using MTR is not banned, are also using it to help affected individuals. However, at the time of writing this book, MRT is effectively banned in the USA, where applications to the FDA for work on human germ cells are not accepted.

There has been little information in the public domain reporting on the heath of the 3 parent babies produced by MRT, as the report will need to cover a whole lifetime. However, the technique has also been used to help infertility, where the mother's oocyte (her egg) is thought to be the source of the problem. The method used for transferring the nDNA of the mother to the donor oocyte is as illustrated for Method 2 in Figure 3.8, and although the mitochondria of the mother are not mutated, the percentage of her mitochondria in the zygote and the developing embryo have been determined[25]. This is possible because the different mitochondrial haplotypes allow sequence differences to be measured. In a couple of these cases, it was found that while only 1% of the mother's mitochondria were present in the early embryo, as the embryo developed, the percentage increased to 70%. This is not important for the new life in these cases, as the mother's mitochondria were healthy (the infertility being attributable to other factors in the cytoplasm of the mother's oocyte). Nevertheless, the data send a warning signal for MRT usage in protecting against mitochondrial disease. The two methods for transferring the nDNA of the mother to the cytoplasm of the donor oocyte (see Figure 3.8) are technically not identical and it is possible that carryover of mitochondria could be different with the different methods. Clearly the use of MRT for eliminating hereditary mitochondrial disease is still a research area and should be used only by Institutes with the full facilities to document the absence of mutated mitochondria in the embryo and the adult.

Having briefly discussed Mitochondrial Disease, which is relatively rare, we move on to the next Chapter to discuss the epigenetic changes which are involved in the very common and multifaceted diseases collectively known as cancer.

Glossary for Chapter 3

Acetyl CoA	This chemical is the substrate used by acetyl transferases to add acetyl groups to nuclear chromatin. It is produced in mitochondria from the pyruvate (derived from glucose or fatty acids) which enters from the cytoplasm.
Aerobic bacteria	Can use oxygen to produce energy.
Anaerobic bacteria	Do not use oxygen to produce energy.
ATP	Adenosine triphosphate is the molecule formed by biochemical reactions in the mitochondria. ATP can store energy for use throughout the body.
Endosymbiotic	The combination of two organisms, one living inside the other with each benefitting from the others presence.
ETC	The electron transport chain: a series of complexes on the inner membrane of the mitochondria carrying and passing electrons to produce ATP and H2O.
Eukaryotes	Multicellular organisms where the cells have a membrane-bound nucleus where their DNA (nDNA) is located in chromatin with histones and nucleosomes.
Haplotype	Individuals who share specific sets of genetic variations in their mtDNA have a common haplotype.
HEFEA	Human Fertilisation and Embryology Authority in UK.
Heteroplasmy	This defines the average percentage of mutated mtDNA compared to non-mutated mtDNA in the mitochondria of an individual who is carrying the mutation. This percentage governs the phenotype off disease developing in the individual.

Mitochondrial Eve	The name given to the individual females who left Africa around 100,000 years ago to give mitochondria to everyone in the world by passing it through the matriarchal lineage.
Mitophagy	Process whereby aberrant mitochondria with mutated genes are auto-digested and destroyed.
MRT	Mitochondrial replacement therapy is the term given to the process of replacing the mitochondria in an egg of a female with mitochondrial disease by mitochondria from a healthy donor.
mtDNA	The DNA in a circular structure carried in the cytoplasm of the mitochondria.
Oxidative phosphorylation	The process used efficiently by the mitochondria to produce ATP to be used by all the cells of the body for energy production.
PAMPS	Pattern-associated molecular patterns are recognised by PRR.
Prokaryote	A free-living unicellular organism: these are bacteria and archaea.
PRR	Pattern recognition receptors in immune cells, they recognise PAMPS in bacteria and other pathogens.
ROS	Reactive oxygen species produced by overactive oxidative phosphorylation which can induce inflammation and damage mtDNA.
SAM	S-adenosyl methionine is the substrate required for methylation of nDNA and histones.

SIRS	The systemic inflammatory response syndrome (SIRS) can be induced after a trauma due to liberation of the contents of the mitochondria which express PAMPS to stimulate the immune response.
TCA cycle	Tri carboxylic acid cycle takes in Acetyl CoA, (produced in the mitochondria) to produce the electron carriers (NADH and FADH2) for the Electron Transport Chain.
$tRNA^{LeuUUR}$	This mitochondrial transfer RNA is responsible for incorporating leucine into a protein by recognising the triplet UUR in the messenger RNA. The gene is mutated in some mitochondrial diseases.

Chapter 4
Epigenetics of Cancer

Cancer the Crab: *a poem*

"Ruled by the Moon, the lady of the night. Mystery binds you to the pale moonlight. You may wear your heart on your sleeve but there's more to you than you would have us believe.
It's not easy to get to know you. But that doesn't mean that it's not worth trying"
Ms Wren

The origin of the word cancer has been attributed to the Greek Physician, Hippocrates (460-370 BC) who used the words *carcinos* and *carcinoma* (which refer to a crab in Greek), to describe tumours. The association of cancer with a crab is inspired by the known ability of cancer cells to metastasize, where the extended metastases can be visualised as finger projections suggesting the shape of a crab. When the Romans came along, the physician Celsus (25 BC - 50 AD), translated the Greek term into *cancer*, the Latin word for crab. Another word still used today is this word oncology which comes from the Greek word for swelling (Onγos), used later by the Greek physician Galen[1].

Life in Balance

As we discussed in Chapter 2 (and is illustrated in Figure 2.2), in a healthy human being, maintaining the balance between cell proliferation and differentiation is crucial for the stability of a tissue or organ but is also highly dynamic. The fully differentiated cell survives and functions for a limited period of time before dying and being replaced by a newly differentiated cell. The successful balance is maintained by the correct functioning of the epigenome in regulating gene expression as the cell divides and progresses to the differentiated cell which does not have the capacity for further division. It is therefore not surprising that in cancer, which is a proliferative disease, the correct balance between proliferation and differentiation is lost, arising from a disturbed functioning of the epigenome of that tissue.

Although mutations in DNA, which change the sequence of a gene play an important role in oncogenesis, the net result is increased proliferation and a disturbed epigenome. Moreover, high rates of mutation are seen in epigenetic factors themselves, with specific mutations being related to specific cancers. More recently, it has been shown, with lower organisms, that cancer can be induced merely by changing the level of expression of an important epigenetic factor such as polycomb (see Chapter 2)[2]. It is important to investigate the changes that occur in the epigenome, as a route to identifying targets for cancer therapy or diagnosis. In this way several therapies have been introduced into the clinic using drugs which target epigenetic factors and some of the epigenetic changes seen in cancer are leading the way to identify novel diagnostic factors.

Since the cancer cells have come from our own cells it hasn't been easy to find what is referred to as "targeted therapy" which kills only the cancer cells, but some have been developed, and the search continues with new therapies continuously being tested. Therapies based on targeting the disturbed epigenome are playing an expanding role in these searches. Also, the search for better diagnostic tools is intense since earlier diagnosis enables successful therapy.

Some Features of Cancer

Different cancers

It is useful to think of cancers as being in two general groups namely:

1. Solid tumours, most of which derive from the epithelial cells which line glands and ducts in organs within our bodies, as well as in skin. The normal cells divide frequently to replenish functional differentiated cells, and such continuous cell division and DNA replication can lead to mutation of genes or aberrant expression of the epigenome, leading to cancer.

2. Cancers which emerge from cells which begin life in the bone marrow and circulate in the blood. These cells we know as the white blood cells or leucocytes, (leuco-white), and an aggressive cancer developing from them is leukaemia. The white blood cells form part of the “hemopoietic system”.

The normal leucocytes include the cells which fight infection, like T cells and B cells (see Chapter 6), and in acting for the immune response, these too can go through many cell divisions and replications of DNA, thus increasing the chance of mutations leading to cancer.

The common solid cancers are carcinomas developing from epithelial cells, like breast, lung and colon, and can in some cases be cured with surgery – that is if they have not set off to invade pastures new, they have not metastasized. Although the cancers derived from blood cells are in circulation, they too can invade other tissues. With these cancers, because it is easy to take blood, samples can be taken following therapy to see if the number of cancer cells are reduced. For solid tumours which are largely internal but have metastasized or need some kind of therapy beyond surgery, it is more difficult to take samples of the cancer to follow the response to therapy and more sophisticated imaging techniques are being developed and the detection of cancer cells or their DNA in blood is looking promising (see below under Improving Early Detection of Cancer).

Most of the clinical trials introducing novel therapies targeting epigenetic factors have been and are tested initially on cancers arising from the white blood cells, although trials are now also being extended to investigate their effectiveness against solid cancers. One factor to be considered is that the inhibitors being used appear to be more selective in targeting cancer cells (than the classical chemotherapy drugs) while not drastically affecting the growth of their normal counterparts. This could be explained if the components being targeted control the expression of only a small numbers of genes which are essential for survival of the cancer cells), while normal cells are equipped with a multi-layered epigenome which can compensate[3,4]

The tumour environment in solid cancers

When, a cancer is established, the cancer environment is not only made up of the cancerous cells but also of other cells, including fibroblasts (the cells that form the structure and the connective tissue of an organ) and cells from the immune system (see Figure 4.1).

Moreover, the cancer cells can secrete factors or interact with the other cells to modify their epigenomes to have a profile that helps the growth of the cancer cells. The cancer-modified cells within the microenvironment can for example secrete growth factors which help the expansion of the cancer cells. One result of this crosstalk is to disable the normal functioning of the immune system which could otherwise reject the cancer cells with "immune surveillance". It is now clear that the cancer cells can induce proteins that inhibit the function of killer T cells, that is those T cells that would destroy the tumour. Some of the inhibitors targeting HDACs (see below) have some effect in combating this effect on the immune response[5].

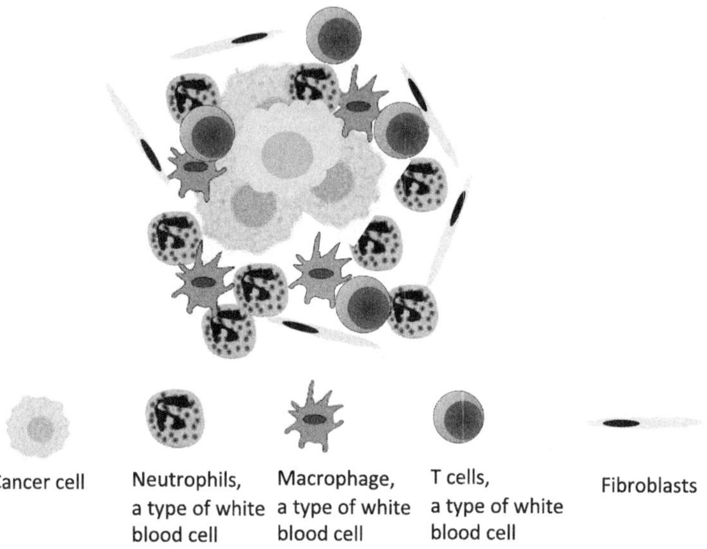

Figure 4.1: The environment around the cancer cells contains a variety of cell types that make up the microenvironment. *Neutrophils are a type of white blood cell and are often the first line attack of the immune response. Macrophages can engulf other cells and particles and present antigens to T cells. Some T cells can help the antibody response of B cells and a different type of T cells known as cytotoxic T cells can kill infected cells and cancer cells. Fibroblasts form the structure of tissues and secrete extracellular proteins like collagen. However, as the cancer grows it produces factors that subvert the normal function of these fibroblasts and other cells in the microenvironment thus helping the cancer to grow.*

Samples taken from a cancer therefore contain all the cells associated with the cancer as well as the cancer cells themselves. Thus, the different cells from the samples can be examined *in vitro* using suitable culture conditions and the cancer cells themselves can be investigated in more detail, in some instances by developing an immortal cell line as referred to in Chapter 2. Recently there has been a move to culturing samples from a tumour as "organoids" where all the different cells are cultured together, allowing the crosstalk to occur, and be analysed. In this type of culture system, cancer cells, which can be obtained during surgery or from biopsies, are cultured with the other cell types in a 3D gel - often including immune cells. This allows the heterogeneity of the original tumour to be maintained. With these studies, it is possible to dissect the cross talk between the different cells in the cancer and find possible targets for therapy of cancer. Moreover, as these organoids reflect the characteristics of the individual patients' tumour, they are an excellent resource for drug screening.

Genes that can Induce Cancer and Genes that can Suppress Cancer

Oncogenes and Proto-oncogenes

Oncogene is the name given to a gene which, when mutated, can cause cancer. When not mutated, the gene is called a proto-oncogene and proto-oncogenes play an important role in normal cells in participating in the control of cell proliferation. Increased expression of several proto-oncogenes is also seen in cancer with different proto-oncogenes being over expressed in different cancers.

The story of oncogenes and proto-oncogenes begins with the study of viruses which target lower organisms such as chickens and rodents (but not humans). There are DNA viruses which code for their genes with DNA and studying DNA viruses led to the identification of tumour suppressor genes (see below). There are other viruses which code for their genes with RNA and it is these RNA viruses which led to the identification of oncogenes and proto-oncogenes. Although some individuals might not like the idea of using toads and birds and mice as models to learn how things work, these models have given a lot of basic information that has proved useful for tackling disease in humans.

Viruses are parasites and need to enter a cell to use some of its facilities to produce more viruses. The first thing an RNA virus needs to do is to reverse transcribe its RNA into DNA and incorporate this DNA into the genome of the infected cell. The viral DNA can then be transcribed along with the cell's DNA to make viral RNA and protein for new viruses to be released from the cell to infect other cells. However, in interacting with the cell DNA, the virus can pick up a gene from the host cell which, when mutated, gives the virus the ability to induce a tumour in an infected animal cell. In this way, more than 100 years ago, a chicken virus (Rous Sarcoma virus or RSV) was identified that had picked up a gene in an infected chicken cell which could induce tumours in chickens. This gene is now called an oncogene. This oncogene was found to be a mutated version of a proto-oncogene which is a normal gene found in the chicken.

This proto-oncogene is evolutionarily conserved, that is it is a gene with a very similar DNA sequence to that found in lower organisms and right through the species to humans. Many other examples of tumour-inducing viruses in rabbits, mice, cats, and non-human primates followed the discovery of the RSV, thus defining other oncogenes, and in each case, they are altered versions of the host proto-oncogene. Therefore, the viruses are carriers of mutated proto-oncogenes that are normally part of the hosts' genome.

Peyton Rous first demonstrated the oncogenic feature of the RSV virus (which was named after him) in 1911, but he didn't receive the Nobel prize until 1966 – the longest time between discovery and prize winning. This delay is almost certainly related to the fact that there was a resurge of interest in RSV in the late

1950s after technological developments in tissue culture allowed the oncogenic transformation to be studied in tissue culture without the use of animals. In fact, the definition of oncogenes and proto-oncogenes *per se*, although made possible by the original discovery of Peyton Rous, came only after technological developments allowed more depth investigations.

Oncogenes have been identified from a few human tumours by transfecting their DNA into a cell line which then became tumorigenic, and these have been found to be the same oncogenes originally identified in the tumour viruses infecting lower organisms. In these cases, a mutation has occurred in the human cancer to convert the proto-oncogene into an oncogene. However, this happens in only a small percentage of cancers, and it is the over expression of a proto-oncogene which is seen in most human cancers.

Proto-oncogenes code for proteins that are involved in the control of cell growth and oncogenes can be classified on the basis of the function of the protein products of their normal counterparts - the proto-oncogenes. These are all proteins which play a role in maintaining the balance between cell proliferation and differentiation.

The increase in the expression of a proto-oncogene is related to specific cancers. For example, the *MYC* proto-oncogene plays a role in the development of cancers developing from the white blood cells (leukaemia, lymphomas and myelomas). The change seen in these cells is an increase in the level of expression of *MYC* (along with other genes which enhance proliferation) and since this is an epigenetic change it is potentially reversible. Moreover, one of the epigenetic inhibitors now in clinical trials which we will discuss can specifically down regulate expression of the *MYC* proto-oncogene in these cancers (BET inhibitors see below)[3,4].

Tumour suppressor genes

Genes which have been termed tumour suppressor genes have an important function in maintaining the integrity of the genome and the epigenome and there is a high rate of inactivating DNA mutations in these protein coding genes in cancer. The cell cycle is a series of events that leads to the division of the cell. The proteins translated from the tumour suppressor genes can affect the cell cycle, blocking progress through the cycle and therefore stopping growth and cell expansion when necessary. Figure 4.2 gives a simplified view of the cell cycle (first worked out in yeast by the Nobel prize winner Paul Nurse). The longest phase, G1, is very active - getting everything ready to move into S phase, where DNA is synthesized. S phase is followed by the G2 phase which prepares for the final act, termed mitosis, i.e. cell division – 2 cells from 1.

The action of the RB and p53 tumour suppressors is necessary for normal cell division. First, in order to make sure time is ripe to move on from G1 to synthesize DNA in S phase (the RB tumour repressor*)* and secondly, to allow time for repair of breaks in DNA occurring in S phase, before moving on to G2 and cell division (the p53 tumour repressor) see Figure 4.2. These genes are highly mutated in cancer often resulting in loss of their function which allows the progression of cells through the cell cycle even when damaged or mutated DNA is present. This leads to the development and progression of cancer. We met *BRCA1* in Chapter 1, which is also a tumour suppressor gene, as part of its function is to repair DNA if a break occurs, and *BRCA1* is particularly important for the mammary gland. Some women are unfortunate enough to be carrying a mutated form of this gene in their genome and have a highly increased risk of developing breast cancer. However, it is more common to find "somatic" mutations in tumour suppressor genes that only occur in the cancer cells.

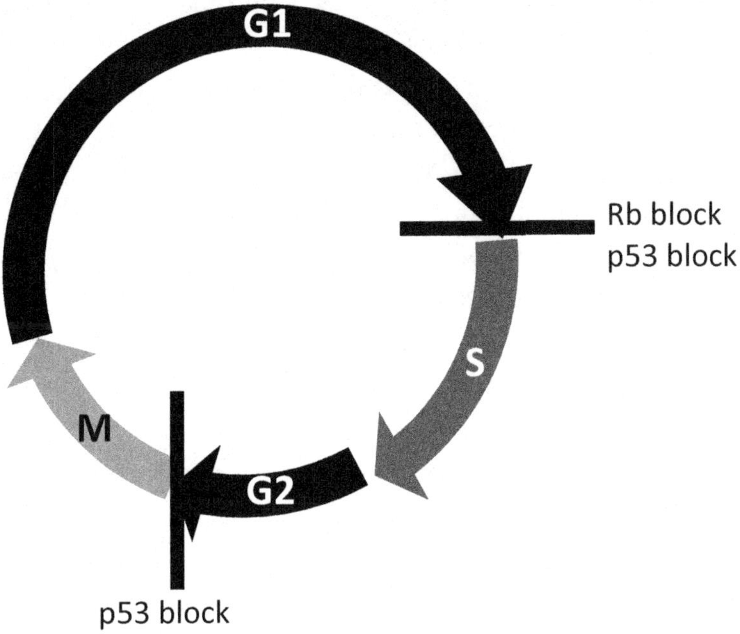

Figure 4.2: Diagram of the cell cycle showing where p53 and Rb block progression. *G1 (gap 1): In this phase the cell carries out a series of reactions and checks to make sure everything is OK to proceed to S phase: (S synthesis) when DNA is copied. Rb acts like the brakes of a car, slowing down progression until everything is ready to move to S phase. In G2 (gap 2), additional checks are applied before proceeding to M (mitosis) where the cell divides to make two cells. p53, the guardian of the genome, is a very important protein, responsible for applying checks to allow DNA repair before proceeding to mitosis.*

The *TP53* gene, coding for p53 is mutated in every type of cancer, the highest rates being in the solid cancers (between 38- 50% of all cancers). The normal p53 protein, the product of the *TP53* gene has been called "the guardian of the genome" as it stops the cell cycle when breaks in DNA occur in S phase and prevents the cell moving into G2 of the cell cycle in order to repair the breaks (see Figure 4.2). In non-cancerous cells, if the breaks aren't properly fixed, the cell then dies by apoptosis, thus eliminating dangerous DNA mutations. The non-mutated p53 in normal cells, also acts as a transcription factor (Chapter 1) and activates the expression of other tumour suppressor genes to help it in its work.

The first encounter with p53 was when it was found bound to a protein found in a DNA tumour virus called SV40[6]. At this stage of the development of cell culture systems, culturing cells from human tissues and cancers was in its infancy and scientists were working with mouse cells cultured *in vitro (in the laboratory)*, particularly mouse embryo fibroblasts (MEFs). Infecting these cells with the SV40 virus "transformed' the cells so that they became immortal and induced tumours when injected into mice. In these experiments, scientists found that a viral protein called the T protein was crucial for the immortalisation and induction of tumours. When they tried to isolate the T protein, it always brought with it a protein of specific size in the transformed cells. (*This could be shown using a technique which had become available for separating proteins according to their size, which, together with charge, determines their rate of movement through a gel*). Their size is reported in a unit known as kilo-Daltons and the protein associating with the T protein was 53 kilodaltons and was therefore designated p53,[7] and the gene coding for it *TP53*.

Later it was shown that in binding p53, the T protein blocked its normal function. Surprisingly when the investigators looked at human cancer cell lines which were immortal, they found high levels of p53, but it was only nine years later, after sequencing the different *TP53* genes, that it was found that the *TP53* gene in each of the cancer cells lines was a mutated form which had lost function.

You can ask - in the light of current knowledge - What took them so long to document that the *TP53* gene was mutated? Well, it's another example of the well-known adage "advances in science depend on new techniques" and sequencing DNA was in its infancy in the 1980s - well before the first human genome was sequenced in 2004. Hence the delay in sequencing the *TP53* gene in the human cancer cell lines. It is likely that the cell lines which were derived from human cancers were immortal because they had a mutation in *TP53,* which, in each case, blocked or perverted the function of the p53 protein. Stopping the function of p53 is clearly a big step for cancer progression, since if DNA breaks are not fixed, the potential for further mutations in other tumour suppressor genes is allowed. Strikingly, mutation of one allele of *TP53* often leads to a deletion of the non-mutated p53 rather than wait for a second mutation to destroy total p53 function.

It is fascinating to note that elephants, who are reported not to develop cancer, have 20 genes coding for p53 (Figure 4.3). Clearly evolution has seen fit to equip them with a great back up so they can effectively ignore single mutations in *TP53*[8].

While mutations are fixed and difficult to target with current technology, the decreases in the expression of non-mutated tumour suppressors which can occur in cancer, are epigenetic and therefore amenable to drug therapy for reactivation. Moreover, re-activation of tumour suppressors has been shown to be induced by inhibiting the function of the factors involved in their down regulation (see below for MLL3 methylase as tumour suppressor).

These early studies with animal viruses identifying oncogenes, proto-oncogenes and tumour suppressors have formed a crucial signpost for understanding the molecular mechanisms involved in human cancer development.

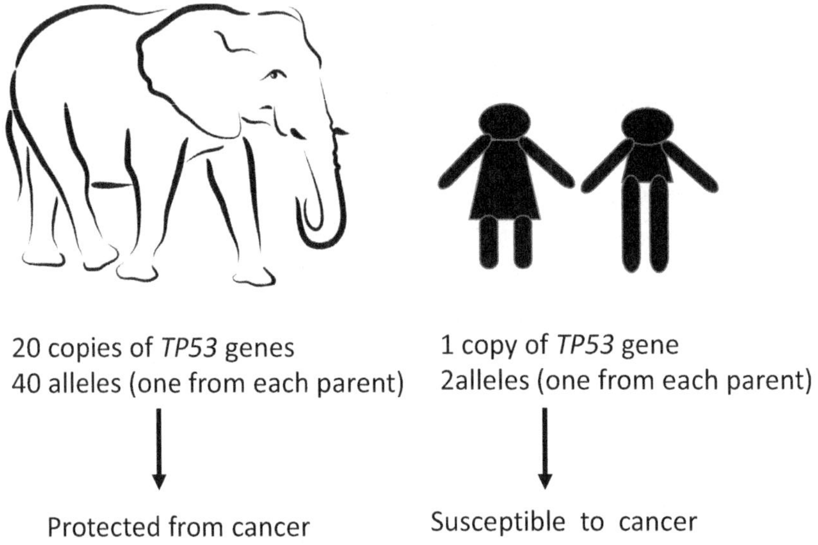

FIGURE 4.3: Elephants have multiple copies of the p53 gene and are reported to not develop cancer*. This provides strong support for the important role of p53 acting as a tumour suppressor. The elephants can accept more than one mutation in TP53 as they have a wonderful back up with so many other functional TP53 genes.*

Can Viruses cause Cancer in Humans?

It is estimated that about 12-20% of cancers world-wide can be attributed to viruses. There are seven different viruses that can cause tumours in humans[9], and some cause very rare cancers and relate to conditions where the immune system is compromised: for example, Kaposi's sarcoma, a rare type of cancer is seen mostly in people with advanced HIV infection (AIDS). However, viruses are not involved in the development of most of the common carcinomas but are connected with the development of specific cancers. Here we will discuss two viruses, the first because it was the first human tumour virus to be identified, and the second because it gives rise to a common cancer in women, i.e. cervical cancer, which is the most common cancer caused by a virus. In cells infected with either of these viruses, the activation of oncogenes and the blocking of the function of tumour suppressors is evident.

Epstein Barr virus (EBV)

The first human tumour virus to be found and studied is the Epstein Barr virus which was found in Burkitt's lymphoma and is associated with other cancers of the white blood cells, nasopharyngeal and gastric carcinoma. Although 90% of the global human population carry the virus, most individuals do not go on to develop an associated cancer. However, a more common result of EBV infection is glandular fever (infectious mononucleosis, IM), which is a self-limiting disease induced by EBV and usually seen in young adults.

The action of the EBV virus not only induces over expression of a proto-oncogene (*MYC*), which plays such a dominant role in cancers of the white blood cells, but also subverts the function of the tumour suppressor gene, *TP53*[10]. The number of new cases for all EBV-related cancers was estimated to be 239,700–357,900 world-wide, with 137,900–208,700 deaths in 2020 but there are no commercial EBV vaccines in existence and no specific EBV anti-viral drugs.

Human Papillomaviruses 16 and 18 (HPV)

There are several strains of the papilloma virus which can infect human cells, but the so-called HPV 16 and 18 are involved in the development of at least four carcinomas - cervical, anal, throat and penile carcinomas. They are the most common sexually transmitted viruses in the world, and cervical cancer is the third most common cancer in women in the world. There is a big difference in the survival of women with cervical cancer in developed nations compared with countries which do not have access to preventive health care measures. In developed nations, strategies have been implemented, firstly to prevent the development of cervical cancer, by offering vaccination to young people (boys and girls aged 12-13 in the UK), and by offering regular screening for women to diagnose the cancer at an early stage. Due to the screening programmes, there has been a decrease of more than 70% in deaths from cervical cancer, from 1955 to 1992 and the vaccination programme in England resulted in a decrease in the incidence of cervical cancer of 97% in those vaccinated at age 12-13[11]. Consequently 90% of the 350,000 deaths from cervical cancer in 2022, occurred in low- and middle-income countries. In these countries where such screening and preventative measures, are not available, cervical cancer is generally diagnosed at an advanced stage when symptoms have already appeared.

So how do HPV viruses cause cancer? They are DNA viruses and upon infection, the viral genome is integrated into the host genome so that important viral proteins can be produced. Two of the viral proteins produced act as viral oncogenes (genes called *E6* and *E7*) which have the ability to decrease the expression of the "guardian of the genome' p53. In this case however, these proteins are also able to disturb the function of the other tumour suppressor- the retinoblastoma protein (RB) which could be termed the "guardian of the cell cycle". Thus, expression of two oncogenes and blocking or perverting the function of two tumour suppressors is a powerful drive for the infected cells to proliferate without its normal control system[12].

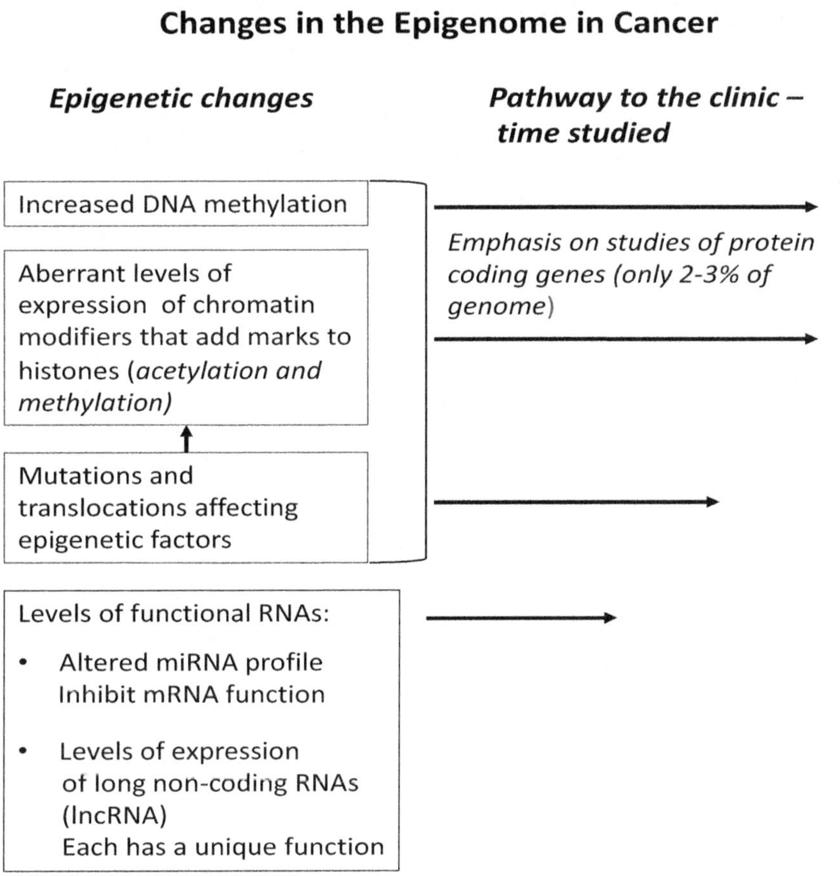

Figure 4.4: Changes in the epigenome occur in cancer. *The changes in DNA methylation and of the enzymatically modifications of histones have been studied for the longest period of time. These studies relate mainly to genes coding for proteins which account for only 2-3% of the genome. The discovery that more than 70% of the genome is transcribed to RNA and that many of these RNAs are functional and show aberrant expression in cancer has led to an expanding study of their role in oncogenesis and a search for new therapeutic targets.*

Epigenetic Changes in Cancer: Can they be Clinically Useful?

Let us look at the changes in the epigenome occurring in cancer and see how this relates to progress in the clinic. Figure 4.4 summarises the epigenetic changes which have been found to occur in cancer and indicates the period of time that the different changes have been studied and investigated for clinical relevance. As indicated in Figure 4.4, studies on the possible clinical relevance of changes in cancer of the methylation of DNA and the methylation and acetylation of histones have been ongoing for many years, while studies on the non-coding RNAs are, as yet, not so advanced. In this context it is important to consider how these studies can lead to increased survival rates for cancer patients. With this in mind, we can consider two main areas where research is being directed:

1. Improving early detection of cancer
2. Improving cancer therapy.

Improving the early detection of cancer

Early detection of cancer can identify cancer at a stage when therapy is more effective and produces less unwanted side effects. In many European countries and in the US, screening programmes such as those for bowel, breast and cervical carcinomas have been introduced with the aim of saving lives. Thus, in the UK for example, there has been a 38% reduction in breast cancer mortality of women being screened[13], and, as described above, an even bigger effect is seen on the development of cervical cancer.

So, what are the changes in the epigenome which can be used for early detection of cancers? Here we can look at two ongoing studies targeting Epigenetic changes which occur in cancer:

- *The COSMOS/microRNA studies* which have been running for many years looking at a high-risk group (smokers) for the early diagnosis of lung cancer.
- *The CCGA /Galleri studies* which represent a more recent development looking for early detection of cancer from many different cancers.

The COSMOS / circulating microRNA study

Since smoking is a strong factor for inducing lung cancer, smokers represent a "high risk" group for searching for a diagnostic tool for early detection of this cancer. Over the last 30 years, in both the USA and Europe, and other developed countries, several groups have been using "continuous observation of smoking subjects" (COSMOS) to try to detect early-stage lung cancer by screening individuals routinely with a scan - low dose computed tomography (LDCT). While cancers have been successfully predictive, whether screening improves survival remains a controversial issue.

Using LDCT as a screening method for early diagnosis of lung cancer is expensive, and there is a high rate of false positives. Moreover, radiation doses need to be considered. Therefore, attempts are being made to look for a liquid screening assay based on profiles of epigenetic factors present in the circulation. The profile of the microRNAs (see Chapter 2) found in the circulation is altered in cancer and the pattern of the miRNA in circulation was selected for further testing to see if this can improve the efficacy and reliability of the LDCT based screening programme.

There are around 2,000 human miRNAs, and by comparing the miRNA profile in the blood of lung cancer patients with that in healthy subjects, the number of miRNAs which are different has been narrowed down to between 20-40 miRNAs, depending on the investigators. This small number of miRNAs is called the microRNA signature classifier (MSC). Several trials have been carried out or are in progress testing the efficacy of screening individual smokers with both LDCT and the MSC. This combined screening is showing a 2-fold increase in the number of cancers predicted and the false positive rate of lung cancer detection is reduced from 20% (LDCT alone) to 3% (LDCT and MSC)[14].

Moreover, the MSC test can predict the aggressiveness of the cancers, with low MSC signal predicting low aggressiveness and showing good survival data. There is also an indication that the diagnosis of lung cancer with plasma miRNA can be detected 2 years before diagnosis with LDCT and these cancers maybe of the more aggressive kind. Thus, a trial has been instigated to explore more fully the details of what can be learned by measuring the miRNA profile sequentially in a large number of smokers and detailing the type of lung cancer which develops in each individual[15].

As with the screening programmes discussed above, these COSMOS studies are investigating screening methods for a specific cancer. However, there is now also progress in developing a screening test for a range of cancers.

The CCGA / Galleria studies

The establishment of a blood screening test for all cancers has been the "holy grail" of cancer diagnosis and it is therefore not surprising that several pharmaceutical companies are attempting to develop such a test. One approach has been to evaluate whether cancer cells undergo a change in the pattern of methylation of their DNA. To be able to distinguish between different cancers the epigenetic change occurring in their DNA methylation would have to be different for each type of cancer. The recent CCGA (Circulating Cell-free Genome Atlas) study is a prospective, observational, longitudinal study designed to characterize the methylation patterns on cell-free DNA in the blood of people with and without cancer. This study enrolled approximately 15,000 participants across 142 sites in the United States and Canada.

DNA is shed from most cells and enters the bloodstream and the CCGA research study analysed the DNA in the blood from cancer patients and from patients without cancer to identify the epigenetic changes occurring in the different cancers. The study was initiated by an International Consortium with the aim to test various DNA based analyses for the detection and localization of a large number of cancer types. Sequencing to define the pattern of DNA methylation out-performed other forms of DNA sequencing detecting mutations, such as insertions and deletions of bases, suggesting that changes in DNA methylation might detect cancer in its early stages. The study was made possible because of the availability of a large data base of the methylation status of various tumour types, allowing the investigators to look at the pattern of DNA methylation found on the circulating DNA and determine a possible cancer signal coming from a specific tissue site.

Clearly these studies require extensive testing in the clinic to see if they are in fact able to diagnose different cancers effectively and specifically early in the disease. Going directly to the public with a test for early cancer diagnosis with individuals who have not been diagnosed with cancer is fraught with the danger of false positives inducing unnecessary anxiety. Therefore, comparing the efficacy of the new test to that of the existing tests for individual cancers (cervical, breast and colon) should have a high priority. In this context, the National Health Service (NHS) in the UK has initiated a large clinical trial (the Galleri trial) in collaboration with the pharmaceutical company GRAIL Bio UK Ltd to evaluate whether using the Galleri test alongside existing cancer screening can help to find cancer early when it is usually easier to treat. Around

144,000 volunteers aged 50-77 who have not been diagnosed or treated for cancer in the last 3 years have registered to take part.

The wisdom of initiating this trial has recently been heavily criticised by cancer experts in the UK[16] not just because of the fear of creating anxiety in the recruited patients, but because earlier studies by the company supporting the trial indicated that early diagnosis of cancer with the new test has been much less effective than expected. It appears that evaluating the use of the DNA methylation parameter for cancer diagnosis would be better investigated at this stage at the research level and on a narrower patient cohort in the clinic. NHS England have decided to wait to see the results of the trial, expected in 2026, before considering whether the NHS rollout of the Galleri multi cancer early detection test should go ahead.

Epigenetic Changes which have led to Development of Cancer Therapies

The epigenetic changes seen in cancer which have led to the development of therapies for cancer are in methylation of DNA and in acetylation or methylation of histones (see diagrams from Chapter 1 and 2). The changes occurring in cancer are summarised in Figure 4.4 along with the time period they have been investigated in the clinic[17].

There are five major epigenetic therapies being evaluated in cancer and of these, four are targeting and inhibiting enzymes which repress expression of genes, including tumour suppressor genes. Each of the four therapies, have the capacity to reactivate these tumour suppressor genes through different pathways (see Figure 4.5). These enzymes in the cancer cells are also inhibiting some of the functions of the host's immune effector cells and of the production and presentation to the immune system of tumour antigens[18].

Two of these therapies have been used in the clinic since the early years of this century and hundreds of clinical trials are currently active with these therapies which use inhibitors of either DNA methylation or of histone deacetylation. Both of these functions are carried out by enzymes- i.e. the DNA methylases and histone deacetylases.

The other two therapies in this category of inhibiting enzymes which down regulate gene expression relate to regulation of methylation of histone tails. These therapies however are at an earlier stage of clinical evaluation. Defining an acceptable non-toxic dose and selecting tumours which show response to therapy does take time.

Targeting DNA methylation

Aberrant levels of methylation of DNA are seen in cancer, and this is being exploited in the Galleri trial, for early diagnosis of cancer (see above). The development of drugs for inhibition of DNA methylation for use in cancer therapy began in the 1960s, and although they began testing in the clinic around 1980, it was only early in the 21st century that permission for use in some cancers was approved by the FDA in the United States (Federal Drugs Administration US). Nevertheless, there are currently hundreds of clinical trials in operation using inhibitors of DNA methylation[18]. These DNA methylase inhibitors have been used mainly for cancers of the hemopoietic system, with a positive effect being seen in patients with the myelodysplastic syndrome (MDS) which is found in the elderly (average age 70-75 years) and can lead to acute myelogenous leukaemia (AML).

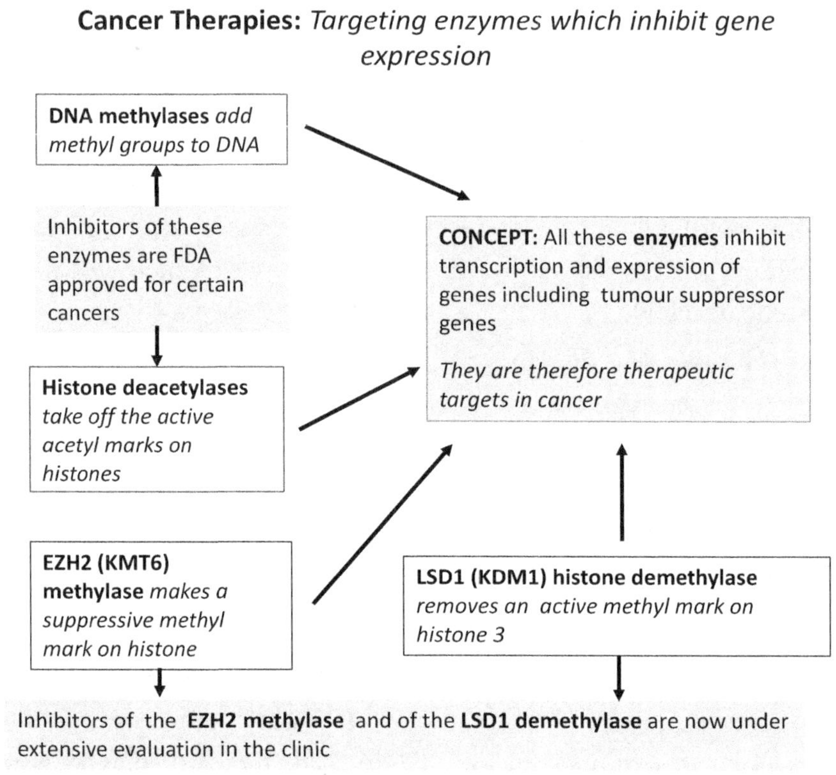

Figure 4.5 Epigenetic therapies in cancer which use inhibitors of enzymes that block expression of some genes, including suppressor genes. *Inhibitors of DNA methylation or of histone deacetylases have been in the clinic for many years and their use is approved for some cancers by the Federal Drugs Administration (FDA) in the US, NICE in the UK and by comparable bodies in other countries. The use of inhibitors affecting the methylation of histones (EZH2 methylase and LSD1 (lysine demethylase 1)) are now under evaluation in many clinical trials. EZH2 was the original name, given before it was realised that it was a methylase, and is still used. It is now also referred to as KMT6. LSD1 (lysine demethylase 1) was the original name of what is now also referred to as KDM1.*

Targeting histone deacetylases (HDACs)

The level of expression of HDACs is increased in cancer and inhibitors of HDACs have been evaluated for cancer therapy since 2006. These HDACs down regulate gene expression by taking off acetyl groups from the histone tails. As we saw in Chapters 1 and 2, acetylation of histones opens up chromatin allowing access for activation of gene expression. Therefore, by inhibiting HDACs that remove acetyl groups, increased gene expression can be achieved. Inhibitors of HDACs have been used widely for cancer therapy and have approval from the FDA[19]. Inhibitors of HDACs are more widely applied to therapy of solid tumours than the DNA methylase inhibitors. They are also widely used in combination - not only with DNA methylase inhibitors - but also with other therapies. It has also been reported that HDAC inhibitors can modulate the immune response and reduce the blood supply to the tumour (called angiogenesis). The high selectivity of killing cancer cells by the HDAC inhibitors can be explained by the "epigenetic vulnerability" of cancer cells, i.e. by inhibiting the expression of genes vital to cancer cell survival but which can be tolerated by normal cells.

Targeting enzymes modifying histone methylation

Both of the other therapies which are at an earlier stage of clinical evaluation than the two discussed above, target factors regulating histone methylation (Figure 4.5). They do however also aim at reactivating genes which have been silenced by the cancer cells (see Figure 4.2). One therapy inhibits a methylase (EZH2) which puts on the repressive polycomb methyl mark and the other targets a demethylase (LSD1) which takes off methyl groups from the trithorax active mark (see Chapter 2 and Figure 4.5). Expression of the EZH2 methylase is increased dramatically in breast and prostate cancer and inhibitors of EZH2 are being tested for the therapy of some advanced carcinomas. Inhibitors of the LSD1 demethylase are mainly being tested on patients who have cancers developing in the hemopoietic system such as MDS and AML. Scores of trials are currently ongoing evaluating the efficacy of targeting these enzymes.

A word about enzymes - what is the function of an enzyme?

An enzyme is a substance produced by a living organism which can catalyse, that is speed up, the rate of a specific biological reaction. It is almost always a protein, coded for by a gene as described in Chapter 1. The enzyme itself is not destroyed by the reaction it is catalysing so it can be used again and again, and the speed of the reaction catalysed is beyond anything achieved by catalysts outside of life. The efficiency of enzyme action was demonstrated by archaeologists working with a dinosaur fossil which still had soft tissue associated with it. All efforts to use known chemical catalysts to dissolve the tissue failed, but half an hour of treatment with the collagenase enzyme, which as its name suggests cuts up collagen, successfully digested the tissue.

The functioning of the cell is totally dependent on the efficacy of a vast number of chemical reactions catalysed by enzymes, each enzyme being structured to work for a specific reaction. Using the purified enzyme-protein and the correct substrate, the reaction can be carried out *in vitro*, thus providing an assay for enzyme activity which can be used to look for an inhibitor of this activity. In this way a very large number of chemicals can be screened to look for inhibitors of the enzyme activity. Consequently, enzymes are favourite targets for scientists trying to identify inhibitors to be tested for cancer therapy.

Shifting Genes about in Leukaemia and BET Inhibitors

Until now, we have discussed therapies which inhibit the function of enzymes which block expression of certain genes, including tumour suppressor genes, and genes affecting immune function. However, the fifth therapy which has been introduced into clinical trials is not inhibiting the function of these enzymes, but instead uses inhibitors to block expression of proto-oncogenes. As we have seen earlier, histones carrying acetyl groups are associated with active genes and a group of proteins (BET proteins) can bind to these acetylated histones and bring with them factors required for effective gene expression. Consequently, inhibitors have now been developed which block the interaction of these BET proteins with the acetyl groups present on histone[3,4]. These BET proteins not only recognise acetylated chromatin which is accessible for transcription factors to activate gene expression, but they also bind to and bring factors in complexes which assist in the process of transcription. In this way the BET proteins can increase the expression of a set of genes which include the proto-oncogene *MYC* and other factors affecting transcription.

The development of the inhibitors of the BET proteins came out of studying what happens in the leukemic cells. In a high percentage of leukaemias, a rearrangement of genes occurs which leads to the production of a new fusion gene which produces a new "fusion protein". This rearrangement is known as a translocation and results from the aberrant repair of two double stranded breaks of DNA occurring on two different chromosomes. In this way a new "fusion gene is formed consisting of two pieces of genetic material from the two different chromosomes fused to each other (see Figure 4.6). This novel gene codes for the fusion protein which can change the profile of genes expressed.

So, in many leukaemias a piece of the novel fusion gene, let's say part I, is always the same and comes from the gene coding for a specific protein (a methylase, KMT2A.) The other piece - part II, which is translocated and fused to Part I, can be different in different cancers. It is however always a gene which codes for a protein which is part of a bigger complex enhancing gene expression (complex B see figure 4.7). Part 1 from KMT2A in the fusion protein can also bind to a different factor in another complex (Complex A) which also enhances transcription, Thus, both parts of the fusion protein bring factors which enhance gene expression.

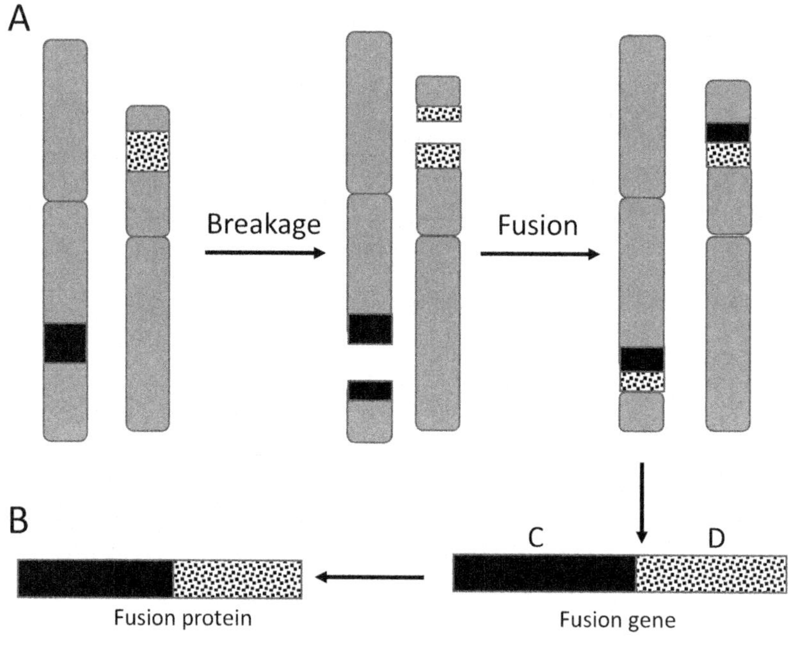

Figure 4.6: Translocations can occur between chromosomes. *In a high percentage of leukaemias, rearrangements of chromosomes can occur, which are known as "translocations".* ***A:*** *DNA in chromosomes can break and, in some cases, when the breaks are repaired, the breaks at two different chromosomes can be joined, rather than being repaired correctly. In this way, a piece of chromatin from one*

chromosome is translocated and fused with another chromosome. **B:** *This translocation results in the formation of a fusion gene made up of 2 parts coming from different chromosomes (C and D). This new gene, in turn, can give rise to a novel fusion protein.*

It turns out that the BET proteins mentioned above also bind to these same A and B complexes so that when they bind to histone acetyl groups, they bring with them the A and B complexes and the fusion protein to an area of chromatin which is open for transcription (see Figure 4.7). The fusion protein can then direct aberrant gene expression, elevating expression of genes which are not normally expressed at high levels. These genes include the *MYC* proto-oncogene as well as other proto-oncogenes and transcription factors.

As shown in Figure 4.7, inhibitors of the BET proteins block their binding to the histone acetyl groups and consequently the fusion protein is not recruited to chromatin and cannot activate the expression of oncogenes.

It has been possible to develop immortal cell lines from individual leukaemias, expressing fusion proteins, where part II of the fusion protein is different. However, in each case, treatment with the BET inhibitors results in decreased expression of the *MYC* proto-oncogene and other transcription factors. This is seen, not only in the leukemic cell lines, but also in lymphoma and myeloma cell lines. The inhibitors blocking the binding of BET proteins to acetyl groups on histones are being widely tested for therapy not only for cancers from the white blood cells but also for some carcinomas.

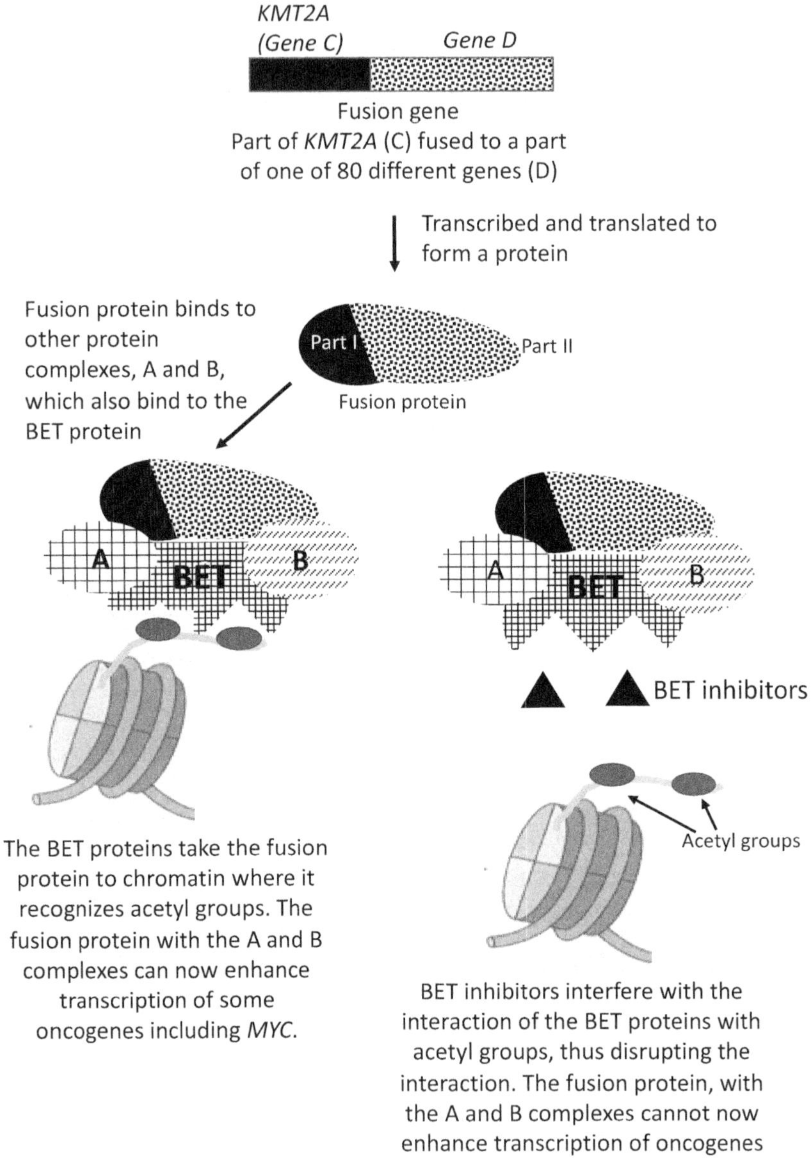

Figure 4.7: Small molecule inhibitors that interfere with epigenetic control of gene expression have potential as anti-cancer drugs*. Shown is how BET inhibitors, that are being investigated in*

clinical trials, can decrease the expression of oncogenes such as MYC in leukaemia and other cancers of the white blood cells.

Mitochondria and possible roles in Cancer Therapy

MYC and mitochondria: A novel target for therapy of MYC expressing cancers

Since *MYC* is an obvious target for cancers over expressing this proto-oncogene, why are there no inhibitors which would inhibit the function of MYC? It has proved difficult to develop inhibitors to target MYC itself, since MYC controls the expression of over 1000 genes and scientists have searched to identify which of these genes are cancer related and are possible targets for cancer therapy. The studies show that genes involved in mitochondrial structure and function are over-represented among the MYC-activated genes, and that MYC plays a major role in mitochondrial biogenesis.

One gene whose expression is controlled by MYC has been identified as a nuclear gene coding for a protein which is responsible for controlling the transcription and replication of mitochondrial DNA (see Chapter 3). This gene codes for a protein, a polymerase (POLRMT) which is not only crucial for the transcription and replication of mtDNA but is also required for the translation of the mitochondrial messenger RNAs into proteins within the mitochondria. This polymerase is very different from other human polymerases which make mRNA from nuclear DNA genes and is most closely related to a virus (T7) which infects bacteria. It is likely that the gene coding for POLRMT (sometimes called *POLGA)* came in with the original bacteria, and over time was transferred to nuclear DNA (nDNA) to become an integral part of the eukaryotic cell.

POLRMT is therefore crucial to mitochondrial function and unexpectedly, it turns out that POLRMT is also crucial for the oncogenic action of MYC. Looking for the action of MYC in cancer cell lines shows that *MYC* expression can enhance cell growth, but if POLRMT is inactivated or deleted, MYC can no longer function oncogenically and the cells die, that is they "apoptose". A small molecular weight inhibitor has therefore been developed which blocks the function of POLRMT and will be tested for therapy of cancers expressing *MYC*, like the leukaemias[20]. Indeed, this type of therapy is being advocated for cancers other than the leukaemias[21].

Helping therapies to overcome resistance: can the bacterial origin of mitochondria bring a surprise?

As we discussed in Chapter 2 (see Figure 2.2), the normal tissue from which a cancer develops is dynamically active. In the normal tissue, a pluripotent stem cell produces dividing cells which give rise to differentiated cells which function for a while and then die. So how do mitochondria behave in this normal functional pathway? Studies following this normal pathway have found that the pluripotent stem cells which give rise to the differentiated cells have few mitochondria, they divide slowly and require less energy from oxidative phosphorylation. As the normal pathway to differentiation develops, and the differentiated cells become active in performing their relative roles, the number of mitochondria increases, the actual number varying with the tissue.

When a cancer develops, it too has a stem cell, and although the production of functional differentiated cells is reduced to allow increased proliferation, some differentiation can occur. (These cancer stem cells are defined as cells that can give rise to the same cancer when grown in isolation). The epigenetic pathways are disturbed in all cancers, but the effect this has on the level of differentiation varies with the cancer as does the number of mitochondria found in the cells at different stages of differentiation. It has been claimed that cancer stem cells have low numbers of mitochondria[22] due to methylation of the gene coding for the POLRMT polymerase.

However, in some cases, the cancer stem cell has increased its use of the energy producing mitochondria by increasing their number. In these cases treatment can be aided by including a therapy which targets the mitochondrial translational mechanism[23]. Because this translational mechanism is the same as that of the bacteria from which the mitochondria developed, antibiotics like *tigecycline,* known to work in this way against bacteria, are being tried in cancer therapy[24]!

In a specific example, treatment of chronic myeloid leukaemia (CML) with a specific inhibitor of an enzyme involved in signalling tumour growth (*Imatinib*) has significantly extended patient survival but resistance develops. It turns out that imatinib primarily targets differentiated cells and does not eliminate the leukemic stem cells (LSCs) which have increased oxidative phosphorylation. Therefore combination-treatment of *imatinib* with the antibiotic *tigecycline,* has been used to selectively eradicate CML stem cells!! There is also a clinical trial in progress treating AML patients with this same antibiotic. We await to see the results.

Can we Target Functional lncRNAs?

The changes which have led to novel epigenetic therapies discussed above have been related to protein coding genes. There are however changes in the levels of expression of non-coding RNAs, referred to briefly in Chapter 2. As indicated in Figure 4.4, studies relating to these changes are not as far advanced in relation to the clinic, as they are with the protein coding genes. However, there are indications that there is potential for these changes to be informative in relation to cancer. Already we have discussed the detection of changed profiles of expression of miRNAs as being studied for the early detection of lung cancers in smokers. Changes in the level of expression of long non-coding RNAs (lncRNAs) are also seen in cancer cells and can enhance cancer progression.

As indicated in Chapter 2 interest was focused on a lncRNA, HOTAIR (HOX antisense intergenic RNA), when it was found to be dramatically increased in several cancer, Moreover, in breast cancer, levels increase as the cells metastasize[25]. In glioblastomas, expression of HOTAIR is activated by a BET protein and inhibitors of BET proteins are effective in slowing down the growth of glioblastoma cells grown in culture[25].

We know that HOTAIR can control the expression of genes on a chromosome different from the one the HOTAIR gene is expressed from: How can it do this? There is one possible clue: Just now we talked about the two factors which are being evaluated as targets for cancer therapy in the clinic, albeit at an earlier stage of testing than the well-established inhibitors of DNA methylation and HDACs. These two factors are, EZH2, a methylase putting on the suppressive polycomb mark and LSD1, a demethylase taking methyl groups off an active mark. Both suppress gene expression, and it turns out that HOTAIR can act as a bridge to bind both of these factors, allowing them to work together at gene promoters to dramatically repress gene expression, including tumour suppressor genes (see Figure 4.8)[26]. Moreover, other lncRNAs have been found to be over expressed in other cancers (gastric cancer and glioblastomas) and they too can act as a bridge for combining the action of the two repressive factors[27].

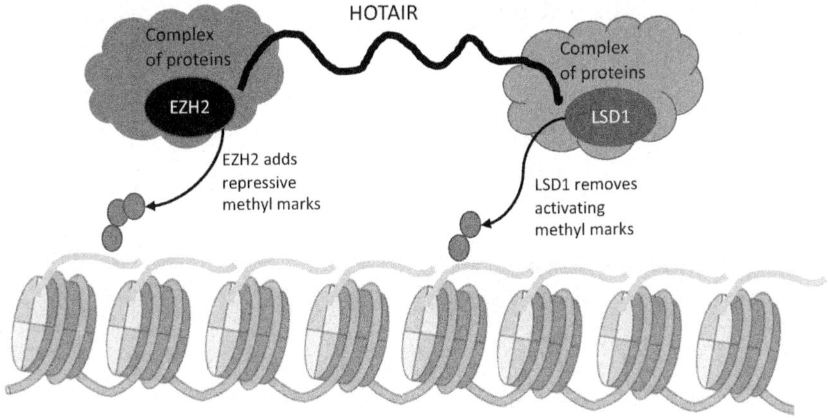

Figure 4.8: Binding of two suppressive factors affecting histone methylation*. Inhibitors of these factors are already being evaluated for cancer therapy in the clinic and could make HOTAIR a potential target for cancer therapy.*

It is now widely thought that inhibiting the function of some lncRNAs could be a therapeutic option for several cancers. However, targeting lncRNA has been a challenge. Recently however, new technologies have been developed and are being tested to meet this challenge. While short stretches of antisense RNAs (siRNA), (i.e. a strand of the bases complementary to the target RNA), have been used to bind to and down regulate expression of lncRNAs, their use is limited because of their sensitivity to enzymes that break down RNA (ribonucleases) and to their being thermally unstable. A novel way of presenting an antisense sequence to block the function of lncRNA has been developed. In this innovation, the bases of the antisense RNA, in the correct sequence, are linked to each other and carried on a more stable peptide type polymer (called PNA or peptide nucleic acid). This polymer is more stable than the antisense RNA. Then just like microRNAs binding to messenger RNAs to inactivate mRNAs, the

PNA carrying the antisense nucleotide sequence can bind to the lncRNA and block its function. Since several lncRNAs have been shown to bind the EZH2 methylase which makes the repressive methylation mark, emphasis has been placed on blocking this interaction, Thus, for HOTAIR, the antisense sequence in the PNA was selected to bind to the sequence in the HOTAIR which is responsible for the interaction with EZH2. Working with cell lines this PNA was found to block the HOTAIR - EZH2 interaction resulting in a reduction in the invasiveness of ovarian and breast cancer cells in culture.

While these are early days for developing methods for the diagnosis and treatment of cancer centred around epigenetics, the recent development of novel technologies for targeting a lncRNA is encouraging. It will be important to harness the lncRNAs for clinical use, where each lncRNA has a unique function and can have a very strong effect on cancer invasiveness. Of course, sorting out the function of the lncRNA could lead to the identification of factors which could also be potential targets, just as the control of expression of HOTAIR by BET proteins in gliomas leads to the use of BET inhibitors appropriate for therapy of these cancers. Since many of the lncRNAs are new genes, only expressed in higher primates, they may very well be important for controlling the stability of the epigenome in complex organisms and their aberrant expression may be involved not only in cancer, but in other diseases, particularly related to brain function.

There have been lots of epidemiological studies suggesting that some factors in the diet can reduce the risk of development of cancer. This is particularly true of the relationship between eating a high consumption of cruciferous vegetables, (such as broccoli and cauliflower) and some cancers. This will be dealt with in Chapter 6 where we will discuss the effect of the environment on changes in the Epigenome. However, one recent approach to inhibiting the development of cancer and to improving cancer therapy is by "starving" the cancer cells. This relies on understanding the differences in the metabolism of cancer cells compared to their normal counterparts[28]. Pre-clinical studies using cancer cell lines and mice have indicated that there may be a difference in the way the cancer cells deal with glucose (which is central to energy production). This can lead to an epigenetic effect since, in cell lines, restricting the glucose in the growth medium resulted in nDNA methylation changes[29]. Moreover, it appears that cancer cells are more vulnerable to a more restricted food availability.

While studies are still largely at the pre-clinical stage, clinical trials have been initiated to test the effect on cancer development of fasting before and after giving adjuvant chemotherapy (chemotherapy given before surgery). Other clinical trials are being initiated to test a more specific dietary restriction such as eliminating from the diet certain amino acids which are required for cancer cell growth[30]. These are early days for directly testing how reducing or altering dietary intake can affect cancer cell growth and the clinical studies are only at the investigative research stage. Since there is currently no quantitative evidence indicating that a restricted diet has benefits for human cancer patients, it is not recommended that patients self-administer a restricted diet.

Observational studies following the health of individuals subjected to famine through war or natural events have been done and these studies will be discussed in the next Chapter. There we will see that some of the data arising from these studies show a surprising effect of diet restriction on longevity and cancer in descendants of the affected individuals.

Glossary for Chapter 4

Angiogenesis	Development of blood supply.
Apoptosis	The death of cells which occurs as a normal and controlled part of an organism's growth or development from the Greek word ἀπόπτωσις, which means falling off.
BET proteins	Proteins which have a domain with which they can bind to acetyl groups on histones in chromatin.
Carcinomas	Cancers developing from epithelial cells. Examples are breast, and some lung cancers.
CCGA	Circulating Cell-free Genome Atlas studying methylation of DNA in blood of cancer patients as marker for cancer diagnosis.
Chaperone	This refers to a protein which is required in order for another protein to function.
COSMOS	Continuous observation of smokers in a screening trial.
Cruciferous vegetables	E.g. broccoli and cauliflower contain sulforaphane (SFT).
Enzyme	A substance (usually a protein) which catalyses a specific biochemical reaction so that the product is produced very quickly, but the enzyme is not changed in anyway.
Epithelial Cells	These cells line the surfaces of skin and glands or ducts.
Fusion gene	A new gene which is made up of components of two different genes fused together and which can produce a "fusion protein" with a novel function.
Haematological system	The cells and factors related to the blood (from the Greek for blood, aima).

Immune surveillance	Refers to how the cells of the immune system survey the cells in the body to recognise cells which are "non-self" (such as pathogens or early cancer cells). System is not working well when cancers or infections develop.
LDCT	Low dose computed tomography used to screen smokers in COSMOS study for early lung cancer.
Leucocytes	The white cells circulating in our blood. Frome the Greek for white (leuko).
Leukaemias	Cancers developing from some of the white blood cells.
Meiosis	The process of a cell dividing to give two cells which only have one chromosome from each of the 23 pairs, as in the formation of an egg or a sperm.
Metastasis	Cancer which has spread from its original site of development to another part of the body.
Mitosis	The process of one cell diving and producing 2 cells, each with the normal pairs of chromosomes
MSC	microRNA signature classifier used to augment LDCT in screening for early lung cancer.
Oncogene	A mutated gene which can cause cancer. Sometimes found in viruses which are oncogenic.
PNA	Peptide nucleic acid where in place of siRNA, the short anti sense RNA sequence is carried on a stable peptide support.
Proto-oncogene	One of our own genes which has a normal function in control of the cell cycle but can be mutated to form and oncogene.

shRNA	A short hairpin RNA is an artificially made RNA that contains a sequence that will bind to a specific RNA and has a tight hairpin turn and is used to inhibit the expression of a messenger RNA (mRNA). shRNA can change the genotype and phenotype of the cell by permanently reducing the level of expression of the target gene.
siRNA	A short RNA with a sequence that can bind to DNA or RNA that has a complementary sequence. siRNA can reduce the expression of a messenger RNA (mRNA) temporarily. The phenotype of the cell does not usually change permanently.
Somatic	Pertaining to events happening to the cells in the body of an animal after birth (from the Greek soma meaning body).
Translocation	The process of moving part of a gene in chromosomal DNA to make a fusion gene with DNA from another gene.
Tumour-suppressor gene	One of our genes whose normal function is to suppress cancer growth, and which is often mutated so as to be non-functional in cancer.

Chapter 5
Are We Hostages to what We Eat or Don't Eat?

"In general, mankind, since the improvement of cookery, eats twice as much as nature requires"

Benjamin Franklin

In the previous chapters we learned something of the mechanics of epigenetics, the complex mechanisms of how cells control which genes are expressed, both protein coding and non-protein coding genes. Certainly, the environment and what we eat play crucial roles in how our body develops and its susceptibility to disease. For example, during early pregnancy folic acid is vital and is often given as a supplement as it helps form the neural tube, and lack of folic acid can lead to defects in the baby's brain and spinal cord such as spina bifida. Folic acid which cannot be made by the body, is involved in the synthesis of the DNA bases and of several amino acids (the building blocks of proteins) and its deficiency plays a role in several other diseases including cardiovascular disease. Another example is that too much exposure to ultraviolet light by excessive sunbathing or use of tanning beds can damage the skin cells and considerably increase the risk of a type of skin cancer called melanoma. Moreover, identical twins, where the siblings share exactly the same sequence of DNA and therefore genes, do not have the same susceptibility as each other to common diseases. Thus, our bodies interact with the environment in many ways.

So, can the environment influence our epigenome and thus influence our well-being in more subtle ways? This is difficult to test in humans as the experiments cannot and perhaps should not be done but we may be able to learn from some catastrophic events that have happened in our relatively recent history by observing what has happened to the individuals experiencing those events and looking at their offspring. This type of research is called "observational studies", that is the scientific community is observing an effect but cannot influence the effect in any way. These studies are important as the results could suggest modifications in human behaviour that may influence human health.

The historical events that have been most studied relate to times of famine, either brought about by a natural disaster or by war. Scientist have observed the consequences of famine in two main ways:

1. They have studied individuals whose mothers, when pregnant were exposed to famine before the individual was born, i.e. while they were still in the womb (*in utero*).
2. They have studied the offspring. of individuals who were exposed to famine during the so-called "slow growth period". This is the time in a person's life just before puberty.

Exposure in the Womb to Nutritional Deprivation

In the 1950s it was believed that the foetus was protected from all nutritional damage that may be inflicted upon the mother, as the placenta was regarded as the perfect filter, only allowing non-harmful substances through. Indeed, in the late 1950s and early 1960s pregnant women were told it was fine to "light up a cigarette and knockback a few drinks" and in 1960 around half of the pregnant women in the USA smoked[1]! Subsequent research has shown how dangerous these activities can be for the unborn baby. A British physician and epidemiologist, Dr David J Baker was one of the most famous champions of the idea that the nine months in the womb is one of the most crucial of an individual's life, shaping a person's adult health and future abilities. This idea led to the studies looking at the health of adult individuals whose mothers were exposed to famine when the individual was *in utero*.

From the late 19th century there have been a number of events, both natural and Man-made that have resulted in severe restrictions in food intake, allowing observational studies. While there are many such events where famine affected a large number of individuals, the collection of relevant data has not always been possible or is fragmentary. However, there is one event which occurred towards the end of the second world war, where the effect of famine on the subsequent health of adults exposed to famine as foetuses or embryos in the womb has been extremely well documented. (Embryo defines the first eight weeks of the developing pregnancy, foetus is from eight weeks till birth).

Dutch Hunger Winter

Towards the end of the second world war, November 1944 to May 1945, the Nazis cut off the food supply to the western part of the Netherlands in retaliation for the exiled Dutch government supporting the Allies. 1944-1945 was an extremely cold winter, the crops had failed, and the population was forced to eat grass and tulip bulbs to survive. The Western part of the Netherlands was densely populated and around twenty thousand people died, and four and a half million people were affected by the direct or indirect effects of the famine, and this period has come to be known as the Dutch Hunger Winter, see Figure 5.1. There are several reasons why this tragic event offers a unique opportunity to study the effect of famine on a population:

1. There is a very well-defined time period as there was a sudden onset, and relief of the famine when the Allies liberated this part of the Netherlands in May 1945.

2. Food availability was registered accurately throughout the famine.
3. Doctors and midwives continued to provide care and keep accurate medical records.

Over 45 scientific papers have been published on the data from the Dutch Hunger Winter and these are reviewed in a recent publication from the University of Amsterdam[2].

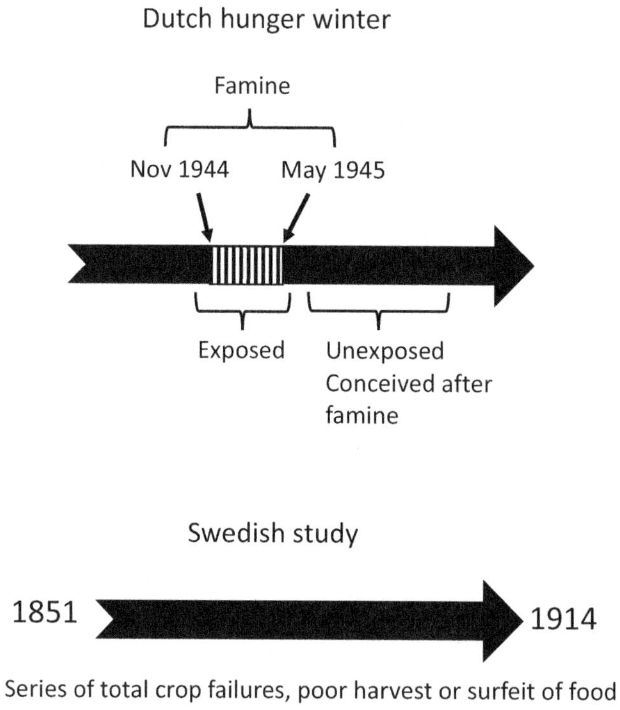

Figure 5.1**: *****Timeline showing exposure to famine in the Dutch hunger winter and the Swedish study.***

The effect on being deprived of nutrients in the womb can be seen on the weights of the babies at birth and on their health in later life and these effects vary, depending on the time of gestation when the mother was exposed to famine. However, individuals exposed to famine early in gestation, that is as embryos, had more pronounced effects in later life. Perhaps this is not unexpected as it is early in gestation when the organs are developing. Embryos, exposed to famine early in gestation, were heavier at birth, while those exposed as foetuses in mid or late gestation were lighter at birth. Women but not men from embryos exposed to the famine were more obese at age 50, and at age 63 had a higher mortality rate compared to unexposed women. However, both men and women exposed as embryos early in gestation had an increased risk of cardiovascular disease and both men and women exposed at any stage of gestation had an increased risk of type II diabetes. Men exposed early in gestation had a higher risk of neurodegenerative diseases, reported more symptoms of anxiety and depression, and had poorer grip strength and lower physical strength. This is confirmation of the theory that the embryo is more susceptible to inadequate nutrition when organs and tissues are developing. At the middle stage of gestation, the bronchial tree grows most rapidly, and embryos exposed to famine at this time of gestation, showed an increased incidence of obstructive airways disease as adults.

So, these were the observations that clearly show that exposure to famine *in utero* has a permanent effect on the individual's health and that the outcome depends on the timing of exposure *in utero*. However, is there any data that suggest what caused these effects and is that epigenetic?

We have learnt in the Chapter 2 that one way of controlling gene expression is the methylation of DNA on the base cytosine in so called CpG doublets found in CpG islands. Individuals who were at an early stage of gestation (ie soon after conception, the subjects) when their mothers experienced the Dutch Hunger Winter, can be compared to their siblings born after the famine (the controls). This was done by taking blood samples from subjects and controls six decades after the exposure and the methylation of areas of the DNA analysed. What was found was that there is differential methylation of areas of DNA between the subjects and the control. Thus 60 years after birth, some areas of DNA showed more methylation in the subjects, some showed less, i.e. the pattern of methylation was different to the controls. Interestingly, some areas of differential methylation were found in pathways related to growth and metabolism[3]. For example, there is a gene known as *IGF2* which is very important in human development, particularly before birth and is a so-called "imprinted gene" (see below in this Chapter). The subjects showed a decrease in the DNA methylation of the *IGF2* gene compared to the control[4], indicating increased expression of *IGF2* which would result in more protein being found in the cell. A word of caution should be stated here. These data do not mean that differential methylation of this *particular* gene is responsible for the outcome observed in the subjects but rather that exposure in uterine to lack of food can affect the methylation of DNA and so alter the epigenome. These data add to growing evidence that early gestation is a crucial period for establishing epigenetic marks that will influence gene expression throughout an individual's life.

In the Dutch Hunger famine, the effects of exposure of the embryo and foetus to famine were only followed in the individuals who developed from the embryo or foetus, and not in their descendants (see Figure 5.2). Thus, in these studies there is no evidence of the inheritance of epigenetic marks. Even if the children of the baby exposed *in utero* were studied this would not provide evidence of inheritance as the baby's germline cells (egg or sperm), the cells that will go on to form the next generation and so the children developing from these cells, will also have been directly exposed. So, to learn anything of epigenetic inheritance the great grandchildren of the woman exposed and so the grandchildren of the baby exposed *in utero* would have to be studied.

If you are a Boy- you are affected by what your Grandfather Ate!!

We have looked at the consequences of exposure to famine *in utero* but what are the effects of exposure during childhood? Here we see the consequences, not only when the children become adults but surprisingly on their children and grandchildren.

The Swedish famine

There is an isolated area of Northern Sweden called Överkalix. In this area, in the 19th and early 20th century, there were acute periods of extreme food shortages due to total crop failure followed by periods of normal harvests where there was a surplus of food (see Figure 5.1). Individuals have been studied that were born in specific years so that the age they were exposed to severe shortage of food is known. What is fascinating is that children who were exposed to famine in what is known as "the slow growth period", this is the time leading up to puberty and is 8-10 years of age for girls and 9-12 for boys, were able to pass the effects of the famine onto subsequent generations. Moreover, the data show that being exposed to famine during the slow growth period more often than not had a positive effect on subsequent generations!

So, what do the analyses show? In fact, the data show some very interesting and surprising results. Some of the earlier analyses indicate that if grandfathers were exposed to severe shortage of food during their slow growth period (SGP), then their grandsons, but not their sons, had a longer life expectancy than the grandsons of grandfathers who had a surfeit of food during their SGP. Now here comes the even more surprising result and that is that this observation was ***only*** seen in the grandsons of paternal grandfathers: not in their granddaughters or in the grandsons of grandmothers[5].

Another study using the same cohort of individuals showed that sons of Fathers who were exposed to restricted food had a lower risk of cardiovascular disease, but this was not seen in the grandsons. However, if the paternal grandfather was exposed to ***excess*** food during his slow growth period, that is the harvests were very plentiful during this period of his growth, then the grandsons had an increased risk of developing diabetes[6]. Yet again, the effects were seen through the male line. In contrast, this study also showed that sons and daughters of mothers who had a plentiful food supply during their SGP had a slight decrease in cardiovascular disease. Very confusing results!!

In 2018 and 2022 further studies were published using data coming from the Uppsala Multigeneration Study which manually recorded and traced the individuals of all births in an Uppsala hospital between 1915-1929. These are known as generation two (G2). Their children could be followed up and are known as G3, and the children's grandparents, known as G1, were traced back from G2 by means of parish registers hospital archives and the Swedish Death Index (see Figure 5.2). This allowed a sample size 40 times larger than the original Överkalix study. Sample size is one of the most important experimental parameters, as looking at small sample sizes, especially with observational studies, may allow results to be seen that are down to chance rather than the effect being investigated.

The 2018 study confirmed the Överkalix data that showed that paternal grandfathers' food access during the slow growth period predicts the longevity of the grandsons. That is grandsons of grandfathers who had a very limited access to food in their slow growth period lived longer than grandsons whose grandfathers had good access to food. Furthermore, this was only seen in grandsons of grandfathers and was not seen in granddaughters of grandfathers or grandmothers. Interestingly, working with larger numbers of individuals, this study could not reproduce the data observed on cardiovascular disease and diabetes seen in the sons and daughters in the Överkalix study. However, what was seen was that the increase in mortality of G3 males whose grandfathers had very good access to food during their slow growth period was mainly due to an increased risk of cancer! Grandsons of grandfathers who had very good access to food had over a 3-fold increased risk in developing cancer than those whose grandfathers had poor access to food during their SGP[7]. A striking and important observation which can be believed because of the large number of individuals examined.

The Turnip Winter 1916-1918

During the first World War, Germany was under blockade by the British and its Allies who from the beginning of the war moved to stop the supply of raw materials and foodstuff to Germany. Much of any food that was available in Germany was sent to the troops on the front line, fighting the war, resulting in severe restriction in the food available for the German civilians. The winter of 1916-1917, later known as the "Turnip Winter", was one of the harshest times as the autumn weather was extremely bad leading to a very poor potato crop and this lack of potatoes lead the German population to rely on turnips coming from Sweden. It is estimated that during this time 424,000 to 763,000 German civilians died of starvation. This famine was well documented and quickly came to an end with the signing of the Treaty of Versailles, making it very suitable for observational studies. In a study looking at mental health, what was found was that the grandsons of paternal grandfathers that were exposed to famine during their slow growth period tended to have higher mental health scores, that is they had better mental health. Maternal grandmothers' exposure to famine also had a positive effect on the mental health of their granddaughters but this relationship of famine in grandparents to good mental health in the grandchildren was not as strong in the female line as the male. What was extremely interesting in this study is that there was no evidence that exposure during the SGP of the parents had any effect on their *children's* mental health. However, as with all observational studies there may be confounding issues to be considered such as the possible lack of reporting of mental health issues during the time period of generation 2 when talking about mental health was often a taboo.

The above discussion opens up many questions including:

1. Why are the observed effects seen more in males then females?
2. Can we explain how the consequences of an environmental exposure, i.e. famine be passed from Grandfather to Grandson?

Firstly, we must emphasize again that all human studies are observational and so it is very difficult to prove a causal relationship. We may be seeing a correlation, for example tall men wear long trousers but long trousers don't cause the men to be tall! It is important to remember this caveat!

Why are the observed effects of Famine Exposure of Prepubertal Children seen more in Male Grandchildren?

Firstly, we must discuss if these traits can be classified as inherited. For an inherited trait resulting from an environmental exposure to be classified as transgenerational, that is truly inherited, it has to be carried beyond the point that the individual that carries it has had direct contact with the stimulus (see Figure 5.2). Thus importantly, the Swedish study showing an effect on the grandsons of individuals exposed during their SFP provides evidence of inheritance of human traits, which is classified as a transgenerational (see Figure 5.2). Could this be the result of the inheritance of epigenetic marks?

When we look at exposure to famine in childhood, during the slow growth period, and the influence on subsequent generations, the effect is much more prominent in boys than girls.

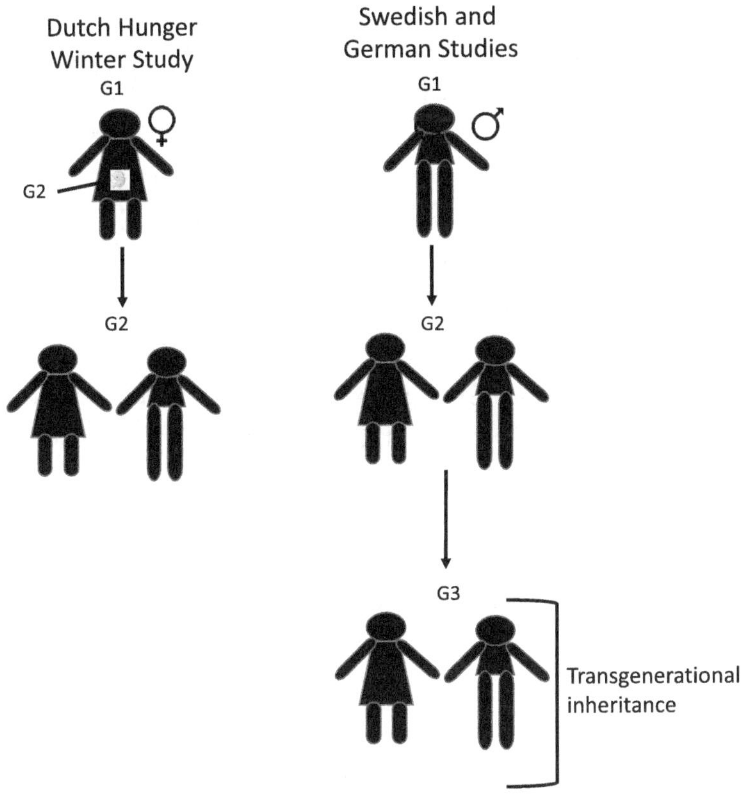

Figure 5.2: Subjects in the Dutch, Swedish and German Studies*. In the Dutch Hunger Winter study, the individual was exposed to the environmental influence, that is famine, while pregnant. Thus, the unborn baby, including their developing gametes (eggs or sperm) will also have been exposed and so studying the adult children developing from the exposed baby does not demonstrate inheritance. In the Swedish and German studies, the individual (grandparent) was exposed as a boy during the slow*

growth period, resulting in their germ cells being exposed. However, the germ cells of their children - G2 - were not exposed and therefore studying the grandchildren (G3) will demonstrate that the traits can be inherited, which is known as transgenerational inheritance.

The effect it is mostly seen when the male parental ancestor is exposed during the slow growth period. This may be due to the different times at which the female and male gametes mature. The development of the female egg is known as oogenesis and begins well before birth. By birth all the eggs that the female will produce are already present in the ovaries. These eggs are known as the primary ova and will not develop further until just prior to ovulation when the egg is released from the ovaries.

Although the primordial male germ cells, that is the cells that will go on to make the differentiated sperm, arise early in embryonic development, spermatogenesis, the development of the sperm, begins at puberty and the slow growth period is a critical time for the DNA methylation of male sperm (see below). During puberty the histones present in the chromatin of the sperm undergo considerable change and most of the histones are replaced by another protein known as protamine. This allows the DNA to be bound even tighter to the protein allowing it to be packaged into the even smaller area of the sperm nuclei. Thus, shocks during the slow growth period of males may have a biological effect on the development of the sperm which they pass on to their immediate descendants. However, in the Swedish study the effect was studied in the grandchildren thus showing transgenerational inheritance.

How can the consequences of an Environmental Exposure, i.e. Famine, be passed from Grandfather to Grandson?

Little research has been published on the inheritance in humans of specific epigenetic marks, for example DNA methylation, histone acetylation and methylation and the profile of ncRNA (Chapter 2), but there is considerable work in animal models, often in mice. However, before we go onto to discuss this, we need to talk a little about how epigenetic marks are kept from generation to generation. The sperm and the egg are highly specialised cells and cannot become anything else: They are at the bottom of the diagram representing Waddington's epigenetic landscape (see Figure 2.1) and have high or moderate levels of DNA methylation respectively. However, when these two cells fuse to form the zygote, this new cell must be able to give rise to all the different cells in the human body, that is it must be pluripotent. Therefore, the DNA methylation that is inhibiting the expression of genes (see Chapter 2) must be removed. Removal of DNA methyl marks on the paternal genome is more rapid than removal from the maternal one, and removal of the marks from both the paternal and maternal DNA takes place in the zygote at the pre-implantation stage of gestation (see Figure 5.3). Moreover, in the zygote, extensive chromatin remodelling is seen, and removal of histone methylation marks are observed[8,9]. There are however some genes coming from either the father or the mother which retain their methylation status, and these are said to be "imprinted genes (see below in this Chapter).

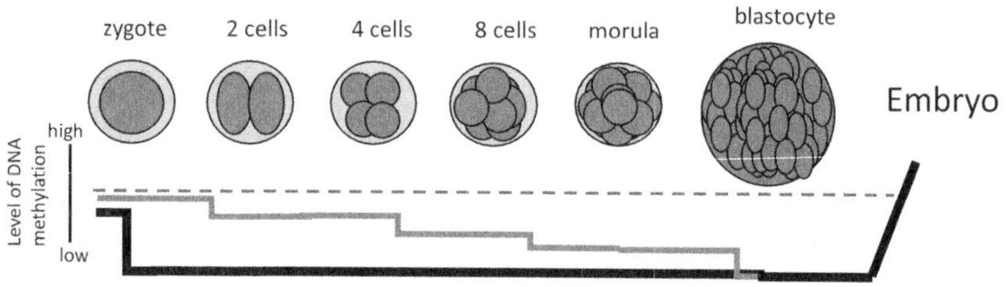

Figure 5.3: After fertilisation the inherited methyl marks on DNA are erased and then re-established in the embryo. *The marks inherited from the father (▭) are erased faster than those inherited from the mother (▬). The marks on imprinted genes (dashed line) are maintained.*

After the blastocyst stage when differentiation begins (see Chapter 2), DNA methylation starts to be re-established by the action of DNA methyl transferase enzymes. Interestingly, some genes harbour both histone repressive and activation marks of gene expression and these regions are called bivalent. This is thought to allow these genes to be held in an inactivated state but to be poised for rapid response to developmental cues. Clearly, as the different organs and tissues are formed, the DNA from different genes will be methylated or activated and histones modified to control the profile of gene expression for each individual tissue and organ.

We need a Mother and a Father: Genetic Imprinting

The egg and the sperm (the gametes) are haploid, that is they carry half the number of chromosomes found in other cells, this means they each carry 23 chromosomes, one of each pair, rather than the 46 seen in other cells. While in females both chromosomes of the sex determining pair are X chromosomes, in males an X chromosome from the mother is associated with a Y chromosome from the father. Therefore, male gametes (the sperm) carry either an

X or a Y chromosome which upon fusion with the egg will determine the sex of the zygote. When the sperm fuses with the egg, the resulting embryo carries the normal 46 pairs of chromosomes, 23 from the father and 23 from the mother. It is diploid. But what if the nuclei from two eggs fused or the nuclei from two sperm fused, there would be the correct number of chromosomes, but would this develop into an embryo? Studies published in the 1980s using mice showed that when this aberrant type of cell with the right number of chromosomes was generated, a viable embryo could NOT be formed. Thus, there is a requirement for chromosomes to come from a male and a female parent to produce a viable embryo, and this is true for all placental mammals. This is perhaps not surprising as the sex determining chromosome pair could be imbalanced in the cell coming from two sperm both containing Y chromosomes, but the sequence of the DNA in the other chromosomes would be virtually the same. However, it turns out that the DNA methylation pattern of some genes in each chromosome pair can be distinguished as coming from the mother or the father, i.e. they are "*genetically imprinted*". For most genes, copies of the gene from both parents are expressed, but for the imprinted genes only one is turned on, either the one from the mother or the one from the father. In this case, the expressed gene is not methylated, and the unexpressed gene is methylated. To allow this, DNA demethylation in the zygote is not absolute and these imprinted genes escape this process (see *dashed line* Figure 5.3).

Thus, not all methylated marks are removed and interestingly the *IGF2* gene we saw to be differentially methylated in subjects exposed to famine as embryos is an imprinted gene. The *IGF2* gene promotes the growth and development of the embryo, by encouraging the growth of the placenta and is normally only expressed from the male chromosome. Here there appears to be a "battle of the sexes" as some genes that promote the efficient functioning of the placenta and so take nutrients away from the mother are often only expressed from the male chromosome. In contrast, genes that are expressed only from the maternal chromosome limit the resources to the baby and so ensure the survival of the mother and an equal sharing of resources to all her offspring. Therefore, most imprinted genes are involved in foetal growth, suckling and nutrient metabolism. This idea is referred to as the "kinship theory" or "parental conflict theory"[10]. In mice there are 150 genes that are imprinted and in humans close to 100[11]. Precise control of the balance of expression of these genes is essential for normal growth and development. Indeed, an upset in the balance of imprinted genes can lead to various health disorders. Although diseases due to this imbalance are rare, the results are often very severe.

However, things get even more complicated because as well as the removal of DNA methylated marks at the zygote stage, demethylation occurs when the sex cells are developing in the embryo! In very early embryonic development, the cells that will go on to make the sperm and the egg are known as primordial germ cells. In a female these cells go onto form the eggs in the foetus and these eggs have half the number of chromosomes, that is in humans 23. In the male, the primordial germ cells in the early foetus give rise to the precursor of the sperm called spermatogonia. These cells still have the diploid number of chromosomes and will not develop into sperm cells, with only 23 sets of chromosomes, until the early growth period of the young boy. These primordial germ cells are the cells that undergo DNA demethylation in the embryo. This erasure is essential to ensure the epigenetic marks can be reset and reflect the sex of the developing embryo[12]. Modifications of histones also undergo widespread changes in the gametes and zygote, although these are not as prevalent as the erosion of the methyl marks on DNA. Moreover, replacement of most of the histones by protamine in the sperm which develop later at puberty, is also a complication for the inheritance of epigenetic marks.

Thus, transgenerational transmission of imprinted marks - or indeed any marks induced by the environment - must escape these reprogramming events. However, it is clear that imprinted genes do escape these measures and therefore it is possible that epigenetic marks induced by environmental conditions can also escape these reprogramming events. However, is another mechanism possible? Several studies have indicated that ncRNAs can be passed to the embryo via their presence in the egg or the sperm and these may well be secondary signals that indicate that certain epigenetic marks should be reinstated. Moreover, recently it has been shown that mitochondrial small-non-coding RNA (mt-sncRNA, see Chapter 3) can be transferred from the sperm to the egg at fertilisation and this can influence the male offspring (see below). At present a lot of this work is being carried out in lower organism including worms[13]. However, these studies are providing increasing evidence that epigenetic information can be inherited across several generations and that these heritable 'epimutations', that is changes in the epigenome, could be part of the process of evolution. Nevertheless, direct proof of this is still not available for our own species, beyond the "observational" studies some of which we have discussed here. So, the mechanism(s) involved in humans in passing on traits to your grandchildren is not yet understood.

Is there Direct Evidence that the Transgenerational Effects observed are due to Epigenetics?

As stated above, in humans there is very little evidence that the transgenerational effects are due to epigenetics. There are however a number of convincing observational studies, in addition to those mentioned above, that demonstrate epigenetic inheritance of traits[14].

For example, about 60% of congenital heart disease that runs in families cannot be attributed to any changes in DNA sequence. Although at present transgenerational inheritance of environmentally induced epigenetic marks cannot be directly linked to changes in the epigenome, there are studies in mice that show a possible direct link. One such study investigated if the fear of a smell could be inherited. Here young male mice were exposed to a particular smell, followed by a startle stimulus[15]. These male mice were then mated to smell naïve females, and their offspring assayed for their response to the smell. These mice were found to have an increased sensitivity to the smell, measured by their behaviour, even though they had never been exposed to the odour. Furthermore, the next generation also showed an increased sensitivity to the smell. These mice, or the sperm and egg that they developed from had never been exposed to the smell, thereby showing transgenerational inheritance.

There are proteins that are involved in odour recognition, and these are known as olfactory receptors. Sperm of the mice that had undergone the smell conditioning showed significantly less DNA methylation of the gene coding for a particular olfactory receptor, when compared to control males, indicating higher expression of this gene. Moreover, when looking at the sperm from the offspring, (this sperm had not been exposed to the stimuli) the same gene showed a pattern of DNA methylation identical to the parent mice. This suggests that this olfactory gene could be more highly expressed in the exposed mice and their offspring, and that a change in the DNA methylation of this gene could be the basis for the inheritance of the olfactory response.

Another recent study[16] mainly using mice as a model looked at paternal diet at the time of conception. Young male mice were exposed to a high fat diet two weeks before mating. 30% of the male offspring from these male mice were shown to be glucose intolerant (measured by high glucose levels after fasting). These glucose intolerant offspring were characterised by having expression of a unique set of genes, particularly those involved in oxidative phosphorylation (see Chapter 3). Interestingly, some of these genes were also found to be expressed in human fat cells and associated with childhood obesity. Using the mouse model, the authors of this paper demonstrated that the mechanism involved in this association was fragments of mitochondrial transfer RNA (mt-tsRNA) found in the male sperm which were increased in the sperm of mice fed on a hight fat diet. As we saw in Chapter 3 mitochondria are passed from generation to generation only through the maternal line. However, it appears that these non-coding mitochondrial RNA fragments can be passed from the sperm to the egg at fertilisation. Importantly, changing the male mice to a low-fat diet for only 4 weeks before mating could reverse the effect in their male offspring. This study shows a direct effect of diet on mitochondria non-coding RNA and how this can affect gene expression in the offspring. There is some evidence that this also occurs in humans as although maternal body mass index (BMI) is strongly associated with offspring, paternal BMI also contributes to the offspring obesity risk. Moreover, analysis of the sperm of young Finnish male volunteers showed that mt-tsRNA is positively associated with BMI.

Can what we Eat Shape our Children and Grandchildren?

So, what do these studies tell us and what can we learn from them? Firstly, there is evidence to support the idea of transgenerational inheritance can occur that does not involve changes in DNA sequence, although the mechanism involved in humans is at present, not understood. Secondly, when pregnant, our diet and environment is extremely important for the health and well-being of our unborn child. A fact that has been known for a long time now. Thirdly, and perhaps more disturbingly is that what we eat as a child just before puberty has an influence on our subsequent generations. Indeed, the Swedish study reported above showed that grandsons of grandfathers who had access to a plentiful supply of food had an increased risk of dying of cancer compared to the grandson of grandfathers who experienced famine. The Western World is suffering a childhood obesity epidemic and the World health organisation reported that 390 million children and adolescences were obese or overweight in 2022[17]. Thus, are we laying the foundations for an increase in cancer in future generations? The evidence from the Swedish study suggests that this may well be the case. However, whatever the consequences on future generations, being overweight is bad for the health of our children.

It is worth noting that we are not too many generations away from the early days of the development of Homo sapiens and from our natural hunter gatherer diet, which was dominated by fresh fruit and vegetables supplemented with some animal protein and nuts. However, the agricultural revolution made it possible to store food and the diet moved to focusing on grains as the main food source – certainly for the poorer members of society. Now, in Europe and North America, selecting food by its taste and its pleasing aspect, and also by its cost, means that pizza, pasta and burgers are not only more generally available but can be less expensive. It is not going to be easy to apply the message of the grandparent to grandson study by persuading the young lads at pre-puberty that greens are good for them and their grandchildren. Somehow it has to seep into a peer group message – as the message for protecting the environment has - and perhaps the message can be passed on in an exciting way in school, early on, while the students still retain their fair share of curiosity?

There is however a growing awareness in some parts of the world (where money and choice of food is operative) of the importance of eating a healthy diet, and of nurturing our immune system and microbiome, which have learned to help us battle negative influences of the environment. In the next chapter we will begin to approach the question of how important the epigenome is in relation to the function of these important systems and indeed to the global health of Homo sapiens.

Glossary for Chapter 5

Blastocyst	The ball of cells that forms in early pregnancy, about 5-6 days after fertilisation. The blastocyst implants into the wall of the uterus and becomes the embryo.
Bivalent genes	Genes that carry both marks for repression and activation of gene expression.
Diploid	A cell carrying two sets of chromosomes one set from each parent. In humans this is 46 chromosomes.
Embryo	Refers to the baby during the first eight weeks of pregnancy.
Epimutations	Changes in the epigenome.
Foetus	Refers to the baby from 8 weeks of pregnancy to birth.
Gametes	The egg or the sperm.
Haploid	A cell containing only one set of chromosomes from the mother or father. In humans this is 23 chromosomes.
Imprinted gene	Where a gene from only one parent is expressed due to selective demethylation of DNA.
In utero	Refers to the prenatal environment of a baby in the womb.
Microbiota	The bacteria which live with us in symbiosis in the gut and other tissues.
Observational studies	Studies where the scientific community is observing an effect over which they have no influence.
Oogenesis	The development of a female egg.
Pluripotent stem cells	Cells that have the ability to give rise to all the cells in the body.

Primordial germ cells	The stem cells that give rise to the sperm or the egg.
Slow growth period	Time leading up to puberty, usually 8-10 years of age for girls and 9-12 for boys.
Spermatogenesis	The development of sperm in boys, the last stages of which occurs during the slow growth period around 9-12 years.
Transgenerational Inheritance	The passing on of a trait to offspring beyond the first generation of children.
Zygote	The multipotent cell formed from the fusion of a sperm with an egg.

Chapter 6
Factors that Influence the Epigenome and Consequently Our Health: What Control do we have?

For ages, bacteria are camping out in human gut.
Nutrition, shelter and love are given them in our cosy hut.
Our relationship is very exquisite
Symbiosis it is called where both are benefitted

Dr Rashi Jain

We have seen that epigenetic marks that control the expression of genes can be inherited from cell to daughter cells and maybe even from generation to generation. However, it is also a dynamic process with enzymes that can add or remove marks from DNA and histones. This influences the expression of RNA that codes for protein (mRNA) or has a function in its own right (non-protein coding RNA). So, if the process is dynamic, can environmental factors influence our epigenome? Indeed, the discussion on the effects of famine in the previous chapter suggest that this may be the case. Furthermore, if we look at the epigenome of the immune system, studies with identical twins, where their DNA sequence is identical, have shown that by the time they are 65, 70% of the immune epigenetics is different, showing that in the immune system much of our epigenome is environmentally determined[1].

It is hard to open a newspaper or magazine without seeing articles about what food is good for you and what is bad, and about how atmospheric pollution is affecting our health but there is often little said about *how* these factors are affecting our health. For example, why is a high fibre diet good for the gut and decreases the risk of colon cancer?

For some time now, sustained efforts have been made to define in detail how environmental factors operating in the life of an individual can affect the epigenome and the health of this individual. Firstly, we should define what is meant by environmental factors in the context of this chapter since these factors can be external or internal. External factors that may affect the epigenome include diet, smoking, stress, physical activity, and these factors are, to some degree, under the control of the individual[2]. There are also external factors in our environment which just happen, and over which we have less control. We saw in Chapter 5 how observational studies on famine induced by war or natural events can have profound effects on the health of exposed individuals and their descendants. In this day and age, and as a consequence of the Industrial Revolution, air pollution in large cities presents a significant threat to health and the control by the individual is limited to pressurising Governments to define legal requirements to reduce such pollution.

In this chapter we will discuss how the food we eat, which we can control, can affect our epigenome. We will also discuss internal factors, which, through diet, we may be also able to control. These internal factors are the millions of microbes that live within our bodies and are collectively referred to as the microbiota.

The Influence of Diet on our Epigenome

In considering the factors over which the individual can have some control, diet is probably the most widely studied. The results of these studies on diet and lifestyle have led to recommendations for healthy eating, and for a physically active lifestyle with some emphasis on the use of some important supplements. In fact, the recommended diet of more fruit and vegetables and plenty of physical activity is very close to the lifestyle which was common to Homo sapiens while living as a hunter gatherer. This phase of our eating behaviour lasted for over 100,000 years to be modified only by farming 5-10,000 years ago.

To show that diet can affect the epigenome, specific changes in DNA methylation, in histone modifications or in non-coding RNA profiles need to be documented when individuals are exposed to different diets (see Chapter 2 for these markers). Consequently, most of the scientific data has emerged from working with animals (mainly mice) to be able to test experimentally the effects of changes in diet on the profile of the epigenome. These studies have shown that what is eaten affects the epigenetic regulation of chromatin by changing enzyme activity for adding or removing methyl groups to DNA or histones and altering the accessibility of factors required for gene expression through acetylation of histones.

In trying to extend these studies to humans, some emphasis has been given to looking at factors in the diet which can affect the expression of histone deacetylases (HDACs) which remove acetyl groups from histones and so can inhibit gene expression. As we saw in Chapters 1 and 2, histone acetylation leads to an opening of chromatin allowing gene expression, and histone deacetylases therefore can inhibit gene expression by removing the acetyl groups allowing the DNA to be move tightly bound to the histones. One inhibitor of HDACs is sulforaphane (SFN, see Figure 6.1), and this substance is found in abundance in cruciferous vegetables, such as broccoli, cauliflower and Brussel sprouts. Broccoli is in fact a relatively cheap and readily available vegetable and a small cohort of individuals were given around a cupful of broccoli to eat and the levels of HDACs and of acetylated histones were then measured in white blood cells. After eating the broccoli, the activity of HDACs was rapidly decreased (3-6 hours later) and the activity of histone acetylation was also rapidly increased in circulating white blood cells. However, this was an extremely small study, with only three individuals involved and no controls of individuals who were not given broccoli, so it is difficult to draw conclusions. Whether this small study suggests that eating lots of broccoli or other cruciferous veg, like broccoli and cauliflowers, can advantageously alter our epigenome is up for debate. However, the data on the effects of SFN on cancer cells are more convincing (see below), so it may be worth including plenty of cruciferous veg in the diet as this is a choice we can make, and it certainly will not do any harm[3].

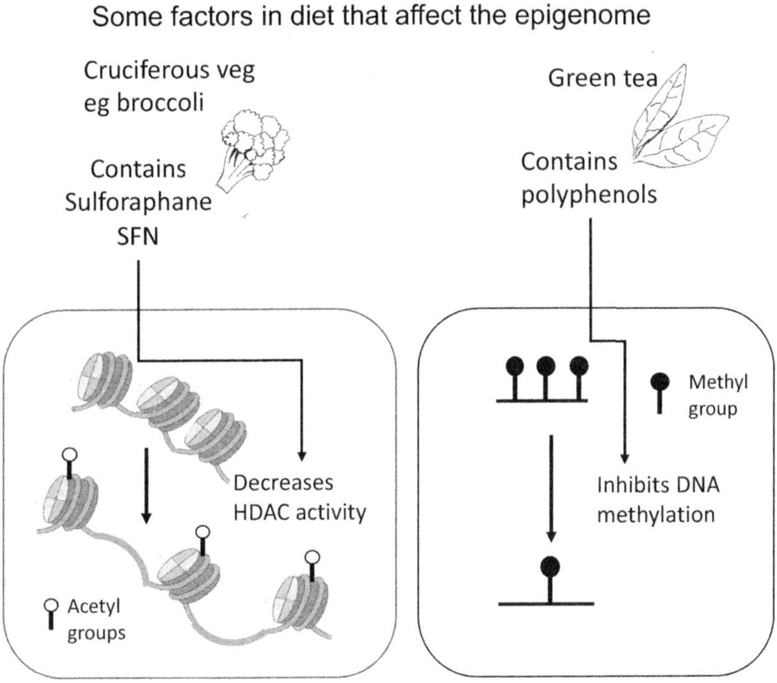

Figure 6.1: Our diet can affect our epigenome*. Sulforaphane (SFN) found in cruciferous vegetables decreases the activity of histone deacetylases (HDACs) which can take acetyl groups off histones and other proteins. This allows the acetylated chromatin to relax and open up allowing the expression of genes. The HDAC6 enzyme can also take off acetyl groups from other proteins and inhibiting its function with SFN disables the function of the Androgen receptor in prostate cancer (see text). Green tea contains*

polyphenols which inhibit DNA methylation and therefore allow genes to be expressed. See text for relevance to prostate cancer.

Can our Diet Impede the Development of Cancer?

There have been lots of epidemiological studies suggesting that some factors in the diet can reduce the risk of the development of cancer. This is particularly true of the relationship between eating a high consumption of cruciferous vegetables and some cancers. Why this might be so has been investigated in some depth with prostate cancer where a mechanism which might explain the connection has been uncovered. The driving force for growth of prostate cancer is the hormone testosterone which stimulates the cancer cells to proliferate by binding to a steroid receptor known as the androgen receptor (AR). AR generates signals which mobilise cell expansion and current therapies aim to disable this growth-stimulatory pathway by blocking the signals from the androgen receptor.

Working with animal models in preclinical studies, scientist have found that sulforaphane (SFN), can inhibit the development of prostate cancer in these models. Moreover, they find not only is SFN acting by blocking the signalling of the AR but they have uncovered how it is doing this[4]. As usual nothing is simple, and this explanation of the action of SFN depends on its effects on a "chaperone" which is a protein that binds to the androgen receptor (AR) to make sure that it functions. The acetylation state of the chaperone determines whether it binds AR or not: When the chaperone is carrying lots of acetyl groups it cannot bind to AR, but when there are few, it can. So, what SFN does is block the enzyme which takes off the acetyl groups so that the chaperone has lots of acetyl groups and cannot bind to AR, which therefore cannot function. Not only this but the level of the AR protein also goes down. We have met the histone deacetylase enzymes (HDACs) that can take off acetyl groups from histones, and some of these can also take off acetyl groups from other proteins. One of these is HDAC6 which can take off the acetyl groups from the chaperone and inhibiting this HDAC with SFN disables it from interacting with the AR receptor. (Being called histone deacetylase is a misnomer as it only weakly takes acetyl groups off histones, but does deacetylate other proteins found in the cytoplasm, including the AR chaperone). As yet there has been no move to evaluate the possible use of SFN for therapy of prostate cancer or any other cancer, but the data coming from several cancers support the idea of including cruciform vegetables, rich in SFN, in our diet.

Green tea has long been thought of as being "good for you". Using a human prostate cancer cell line, it was shown that substances in green tea called polyphenols inhibited the action of an enzyme that adds the inhibitory methyl groups to DNA and increased the acetylation of histones which allows gene expression. There is a gene known as *GSTP1* that is often not expressed in prostate cancer and the silencing of this gene is a hallmark of prostate cancer. However, the scientists found that treatment of a prostate cancer cell line with polyphenol resulted in this gene being expressed in the cells. Importantly, this increase in expression was not found for other genes, where increase in levels of some proteins could be harmful to the cell[5].

Our Microbiota: we can't live without them!

Over millions of years of evolution microbes have co-evolved with hosts in a symbiotic relationship. That is both the host, and the microbes benefit from the association. We have examined one example of this in Chapter 3 where we describe how bacteria entered the eukaryotic cell many millions of years ago to live in *endosymbiosis* as mitochondria within our cells, providing their oxidative phosphorylation system to produce the energy we need. There is however a further important example of *symbiosis* of our cells working with bacteria where microbes, collectively called the microbiota, can live within our tissues and influence the epigenome. In humans, organs such as the skin, uterus, lungs and gut have resident microbes. Over 70% of the microbiota live in the gastro-intestinal tract where there are between 10^{13} and 10^{14} microbes and the gut microbiota has been estimated to weigh 1.5-2Kgs[6]! Since a high percentage of the microbiota in our bodies are in the gut, the gut microbiota have been most widely studied. They play a vital role in human health and to some degree can be under our control as their composition and function can be affected by our diet and can be modified by medication use, diseases, social networks, genetics, and age.

Living without microbes at first sounds like a great idea! No more nasty diseases such as TB or tetanus, no more moulds on the bathroom tiles, no more stomach bugs. However, microbes play a vital role in maintaining the eco-balance not only in the world as a whole but within each and every one of us. The co-evolution of host and the microbiota indicates that the interaction is important to both the host and the microorganisms and indeed the term HOLOBIONT has been conned by some scientists which describes the host and microbiota as a single and evolutionary unit. The microbiota are vital for our general health and for the prevention of some diseases as they have an enormous influence on our immune system.

In humans there are as many microbial cells as human cells. Every micro-organism that co-exists with us has genetic material and this is referred to as the microbiome. This composite genome of the microbial community is in fact 100-150 times larger than that of the human genome[7]. Moreover, the composition and function of the microbiota are affected by diet and other external factors, making it possible, following research findings, to learn about the possibility of having some control over maintaining a healthy microbiome and therefore healthy epigenomes in our tissues.

The microbiota consists of bacteria, fungi, archaea (an intermediate form between bacteria and multi-cellular organisms) and viruses, with bacteria forming the dominant species. 99% of the microorganisms in the gut are bacteria and the strains of bacteria which exist within the microbiota are well documented[8]. Moreover, the microbiota are responsible for synthesizing an abundance of factors that are involved in the epigenetic control of human genes (see Figure 6.2).

It is now abundantly clear that the microbiota are intrinsically and crucially involved in working with our own cells to achieve healthily balanced epigenomes. Moreover, it is also clear that some modifications of the microbiome, occurring in response to external factors are directly involved in the development of several diseases. These include some metabolic diseases such as inflammatory bowel disease (IBD), obesity and diabetes, as well as cancer and conditions resulting from an imbalanced immune system. In the next Chapter, we will select some of these conditions to discuss in more detail how the involvement of the microbiome affects our epigenome. Meanwhile as a background to these discussions, some general information about how the microbiota can influence the epigenome could be useful.

The largest complement of microbes is found within our gut, and the gut microbiota maintains the ecosystem of the human digestive tract. The microbiota in the gut can influence the epigenome in several ways and one important way is by producing molecules which can directly affect our epigenome, see Figure 6.2. As we saw in Chapter 2, methylation of DNA, and methylation and acetylation of histones, play important roles in regulating the gene expression profile of the epigenomes found in the different tissues. In this context, the microbiome can produce the substrates used for these enzymatic reactions and this is illustrated in Figure 6.2. It will not have escaped your notice that we reported in Chapter 3 that mitochondria, which developed from bacteria which invaded cells and are working with the cell, can also supply these substrates. However, these mitochondria interact with us as endosymbiotic visitors, while the bacteria which form the microbiota are living, as bacteria, alongside our cells, secreting these factors, making them more widely available (see below). Bacteria within the gut can produce the chemical known as S-adenosylmethionine (SAM), which provides the methyl groups when DNA and histones are methylated, leading to the regulation of gene expression. Bacteria within our gut can also produce folate from which SAM can be derived and folate is such an important factor for health and for the prevention of spinal conditions in babies that in the United States it is added to bread, cereals and flour and in the UK it is added to non-wholemeal flour. Folate is also recommended as a supplement for pregnant women.

Acetylation of histones (with histone acetylases, HATS) relaxes chromatin to allow gene expression, and requires the substrate acetyl CoA. Short chain fatty acids (SCFA) are produced by the microbiota and SCFA can form a substrate for the production of acetyl CoA by the TCA cycle within the mitochondria (see Figure 6.2). Not happy with just encouraging the HATs, to do their job, the SCFAs among other things, block the function of histone deacetylases (HDACs) thus allowing the histones to keep their acetyl groups, favouring gene expression (see Figure 6.2). SCFA are the most abundant and highly studied of the microbial produced substances and are responsible for using 10% of the calories required by the human body to function efficiently.[9] These short chain fatty acids have many effects on our biology, and these include modulating the level of glucose in the blood, inhibiting excess cholesterol (through the direct action on the enzymes involved in its production), influencing the appetite, acting as an energy source and having anti-carcinogenic and anti-inflammatory effects.

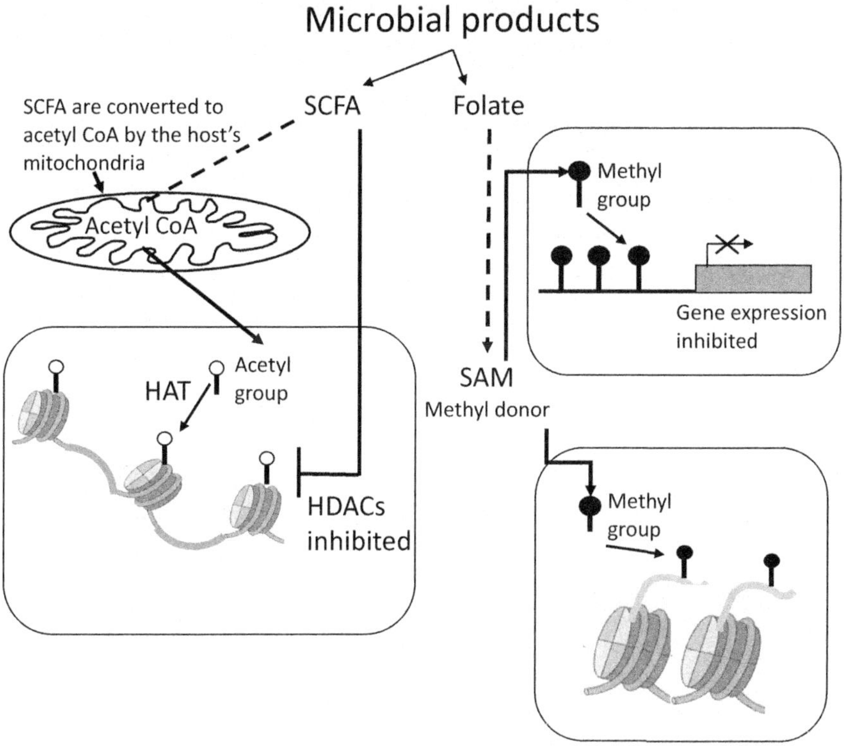

Figure 6.2. Bacteria in the microbiota in the gut produce substances that are required for enzymic modification of our epigenome. *Short chain fatty acids (SCFA) can inhibit histone deacetylases (HDACs) and via acetyl CoA, produced in the host cell's mitochondria, enhance histone acetylase transferase activity (HAT). S-adenosylmethionine (SAM) can donate the methyl groups required for the methylation of histones and DNA.*

Different microbes produce different types and different quantities of SCFA, but one important SCFA component, produced by our gut microbiota is butyrate, which is particularly effective at inhibiting the action of histone deacetylases (HDACs see Chapter 2, 3 and 4). This relaxing or opening up of the DNA allows for the access of transcription factors leading to gene expression (see

Chapters 1 and 2). Thus, butyrate produced by the microbiome can epigenetically induce gene expression changes.

A Useful Model from a Fellow Mammal

The importance of the molecules produced by the microbiota to our general health and in affecting the epigenome is clearly demonstrated by work coming from the study of a mouse model. Much of the data relating to the importance of the gut microbiota, and the products they produce comes from working with germ free (GF) mice that have been reared in a germfree facility to have no microbiota. So, by comparing the modifications of the epigenome of GF mice with those seen in mice carrying microbiota (reared on a healthy diet), conclusions can be made concerning the importance of the microbiota for a healthy epigenome. These studies have confirmed the importance of SCFA produced by the microbiota. Thus, scientists have found that by adding SCFA to the diet of the GF mice, the mice develop a healthy epigenome, similar in histone methylation and acetylation to the mice raised with a healthy diet and a full complement of microbiota.

In these mouse models it is also possible to check the healthiness of the microbiota donated by mice reared on different diets – by looking at their effect on the epigenome. This can be done because it is possible to transfer the microbiota in the faeces from one mouse to another, and microbiota from mice reared on different diets can be transferred to the GF mice that have no microbiota and see how the epigenome fares. This is done by transferring the microbiota in the faeces of the mice to the GF mice (termed faecal microbiota transplant or FMT). First, the scientists looked to see what happened to the epigenome in different tissues when they transferred microbiota from a mouse reared on a polysaccharide rich healthy diet to the GF mice. They looked at the acetylation and methylation of the histones in the chromatin of these tissues and found changes in these marks on specific lysines which were not seen in the GF mice without microbiota[10]. This shows directly that the gut microbiota modifies the epigenome not only of their neighbouring epithelial cells in the colon, but also of other tissues including liver and some adipose tissue. Just how this happens, whether through the circulation or not, requires further research, but it does provide a method for checking how functional specific microbiota are in making these marks when they are transferred from mice raised on different diets. The diet they wanted to test was the "high fat, high sugar diet (HfHs)" eaten by some of us (but not all) in the Western world. Perhaps not surprisingly, they found that microbiota taken from mice which had been raised on a HfHs diet were very inefficient in modifying the epigenome. Not only that, the mice with the HfHs microbiota were overweight, produced much less of the important factors which we ourselves need (SCFAs), but did induce higher levels of cholesterol and triglycerides, high levels of which can be harmful.

Of course, we are not mice, and clinical studies have begun to investigate in the human setting how transplantation of faecal gut microbiota from a healthy individual to a patient with disease could be therapeutic[11]. In this context, in a clinical trial, faecal microbiome transplantation (FMT) using colonoscopy proved to be a very effective treatment for a harmful bacterium that causes diarrhoea, with patients showing 90% cure as compared to 26% cure in patients given an antibiotic[12]. Infection with these harmful bacteria is most commonly found in individuals, especially in the elderly, who have had extensive antibiotic treatment. These trials are with small numbers of individuals and of course do not present data to show the effects of the microbiota are operating through changes in the epigenome. Strikingly FMT was first tested in China in the 4th century where human faecal suspensions were delivered by mouth to treat cases of diarrhoea[13]!! Colonoscopy seems like a more user friendly and acceptable way of transplanting the microbiota and therefore likely to attract more patients to trials! The transfer of healthy microbiota for combating disease, recalls the transfer of healthy mitochondria for the prevention of a hereditary mitochondrial disease.

We need to feed the microbiota

It is very impressive to see how effective the symbiosis is, which goes on between our cells, the gut microbiota and our diet. In our diet we bring in amino acids like methionine which we produce by digestion of proteins and from which the microbiota can produce SAM for DNA and histone methylation. However, since the SCFA they produce, which includes butyrate (see Figure 6.2) are supremely important for the maintenance of a healthy microbiome it is very important that we include in our diet the molecules which they can use to produce the crucial SCFA.

In order for the microbiota to function well, we have to feed them well

So, what is the food we need to eat for them? The components they need are dietary fibres which are "carbohydrate polymers with ten or more monomeric units", which are neither digested nor absorbed by the cells of the human small intestine[14]. They are polysaccharides, a name which reflects the fact that they are made up of many sugar molecules joined together. However, the specific sugar and the linkage can be different. There are polysaccharides like starch where our own cells can cut the linkage between the sugars and starch is therefore digested. The soluble fibres, which we cannot digest but the microbiota can, are required by them to produce SCFA and are found in plant-based food. Those found in grains, fruits, vegetables, nuts, and legumes can be different, so a diet consuming several sources ensures enough food for a healthy microbiota. Fibre which is less soluble, such as cellulose is also present in some plant foods, namely cereal grains and the cell walls of fruit and vegetables. Although cellulose is not digested by mammals and is poorly fermented by our gut bacteria, work in mouse models shows that including cellulose in the diet can alter the composition of the gut microbiota and can protect against colitis and some forms of auto immune disease. Moreover, the modification of the composition of the gut microbiota appears to be a key factor in the maintenance of gut homeostasis. An added effect is that by absorbing water and adding bulk to stools, dietary cellulose can also have a mild laxative effect. Taken together the available data indicate that including lots of cellulose in the diet is good for us even though we are not ruminants.

We have discussed the gut microbiota and most of the work demonstrating the effect of their functions in modifying the epigenome has been done with these microorganisms, but there are collections of microorganisms found in other tissues, for example in skin and in the vagina, and trials are being instigated to examine whether transplanting these tissue-specific microbiota can be therapeutic for diseases found in those tissues. To make more detailed progress with studying Homo sapiens, our microbiota and our symbiotic relationship with it, dictates that we need to consider their needs as well as our own. Because it is becoming clear that malfunctioning of the microbiota is related to various diseases, it is with researching these diseases we may come up with data which will not only help patients but will bring new data in the understanding of the microbiome and the epigenome. In the next chapter we will expand on this by discussing the importance of a healthy microbiome in combating inflammatory diseases of the bowel and how the effect of the microbiota on the immune system plays a central role in cancer and in the response of patients to immune based cancer therapy.

Glossary for Chapter 6

Acetyl CoA	A precursor for the TCA cycle where it is oxidised for energy production.
Androgen	A group of steroid hormones, including testosterone that bind to the androgen receptor.
Archaea	An intermediate form of life between bacteria and multi-cellular organisms.
Butyrate	A short chain fatty acid produced by microbe fermentation.
Cellulose	The main constituent of plant walls and vegetable fibres.
Chaperone	Biological chaperones assist in the correct folding and assembly of proteins.
Cruciferous vegetables	Vegetables from the brassica family such as broccoli, Brussels sprouts, cabbage, cauliflower and kale.
Endosymbiosis	Two organisms living together with one inside the other. In the case of the origin of mitochondria, bacteria invaded the cell.
Germ-free mice (GF)	Mice with no microbiota.
GSTP1 gene	A gene coding for a protein that metabolises a variety of carcinogenic compounds and protects cells against DNA damage. It is not usually expressed in prostate cancer.
HfHs	High fat high sugar diet.
Holobiont	The host and the microbiota living as a single entity.
Microbiome	The genetic material of the organisms (the microbiota) living in a particular environment.

Microbiota	The range of micro-organisms (bacteria, fungi and archaea) living within a particular habitat such as the gut.
Polyphenols	Naturally occurring organic compounds that are found abundantly in plants, especially in green tea.
Polysaccharides	Carbohydrates which consist of a number of sugar molecules bound together (eg starch or cellulose).
S-adenosylmethionine (SAM)	A compound produced by the microbiota that is the main source of methyl groups for the methylation of DNA and histones.
Short chain fatty acids (SCFA)	SCFA are a subset of fatty acids produced by the microbiota by the digestion of food such as polysaccharides and fibre that we cannot digest.
Steroid receptor	A protein found in cells that can bind steroids such as androgens and oestrogen.
Sulforaphane (SFN)	A natural plant compound found in cruciferous vegetables.
Symbiosis	Two different organisms living together for mutual benefit.
TCA cycle	Is a series of chemical reactions that occurs in the mitochondria to help to generate energy.
Testosterone	A steroid hormone that stimulates the development of male sexual organs. Found in the testes in males and in females in a number of tissues including the ovaries, adrenal glands and fat cells.

Chapter 7
The Intimate Relationship Between Microbiota and The Immune Response: Impact on the Therapy of Some Human Diseases

"Like pieces of a puzzle, our differences fit together, in this grand mosaic called life.
Each shape unique, each color vibrant, woven with threads of diversity"
Jeanine Dejesus

We have seen that the microbes within our gut can regulate expression of our genes, so could they therefore influence diseases and human health? In this context, there is considerable data showing that the microbiota can influence the immune response which is there to distinguish self from non-self. 'Immune surveillance' (IS) is the term used to describe the function of the immune system to detect and eradicate abnormal cells: for example, cells infected with a virus can be seen as non-self, and cancer cells that carry aberrant antigens can potentially also be recognised as being "non-self". If effective, IS can act as a protective defence mechanism, but it is regularly subverted by the weapons wielded by the abnormal cells. It is therefore a reasonable question to ask if the quality of our microbiota can function to enhance or to oppose the function of the immune system in combating human disease.

Microbiota and Our Immune Response

Colonisation by the microbiota is crucial for the development of the immune system[1] and the effect of the microbiome on the immune system can impact on the development of diseases such as inflammatory bowel disease and cancer. Although there are correlative and observational studies in humans, by necessity much of the work showing a connection with the microbiome, the epigenome and disease has been carried out in mouse models.

As we have seen the microbiota are made up of bacteria, viruses and other micro-organisms, which are all foreign to us as humans. So, why is it that we do not mount an immune response to these organisms? As we know our immune system is there to fight foreign organism so what is special about the microbiota? As discussed earlier, humans and their microbiota have evolved over millions of years and it is clear that maintaining the correct balance between our microbiota and ourselves is crucial to our immune response and to the prevention of diseases such as inflammatory bowel disease[2]. The cells lining the intestines secrete a substance known as mucus that forms a physical barrier between the underlying cells and the microbiota. However, the microbiota regulate the integrity of this mucus barrier and host immune cells can be found within it. Moreover, it is known that the microbiota within our gut have the ability to alter the activity of the host's enzymes that can modify our chromatin resulting in a modulation of immune activity.

In mice it is possible to study the effect of no microbiota or specific elements of the microbiota on the development of the immune system and this shows the importance of the microbiota in the development of immunity. Germ-free mice (these have no microbiota) have many defects in their immune system. These mice are more susceptible to infection and have fewer white blood cells than pathogen free mice (PF mice). PF mice of the same strain maintain their good microbiota but do not harbour any pathogens and are good for comparing the effect of these healthy microbiota with the germ-free mice. The development of the small intestines is also defective in germ free mice[2] and germ-free mice raised without a microbiota show considerable disruption of the host's chromatin in the intestines[3]. However, these data still beg the question "why don't we reject these organisms"? This of course is a very important question. In a healthy organism with healthy microbiota, the immune system has learned to regard these micro-organisms as "self" and does not reject them. This is related to the question how does the immune system recognise "self "and what happens when this self-recognition process goes wrong, as for example in inflammatory bowel disease? Clearly as with most systems in nature, the regulation of the immune system is dynamic, and the immune response can be dampened as well as activated in order to achieve a healthy balance. While the immune system is very complex, some discussion of important points might be helpful.

The two types of immune response: Innate and Adaptive Immunity

The immune system consists of two arms known as the innate and adaptive immune systems. The cells of the innate immune system are the body's first line of defence against infection and are exemplified by cells such as macrophages (see Figure 7.1), and some soluble factors which can activate cells to start rejecting invading micro-organisms. All animals have some form of innate immunity, but the development of adaptive immunity came later in evolution, about 500 million years ago, in jawed fish[4]. The cells of the adaptive immune system are of two main types namely, those which produce antibodies which can target and kill pathogens like bacteria, or cells which themselves can directly kill cells infected with a pathogen like a virus. What makes these different from the innate cells is that they have "memory", that is - once the body has seen a foreign substance, these adaptive immune cells can remember they have seen it before and can respond to it quickly and with more intensity on the second and subsequent exposures. This is the mechanism behind vaccines where we are warned, usually by the injection of a variant of a pathogen which is not infectious but can alert the adaptive immune cells, to be ready for any future infection.

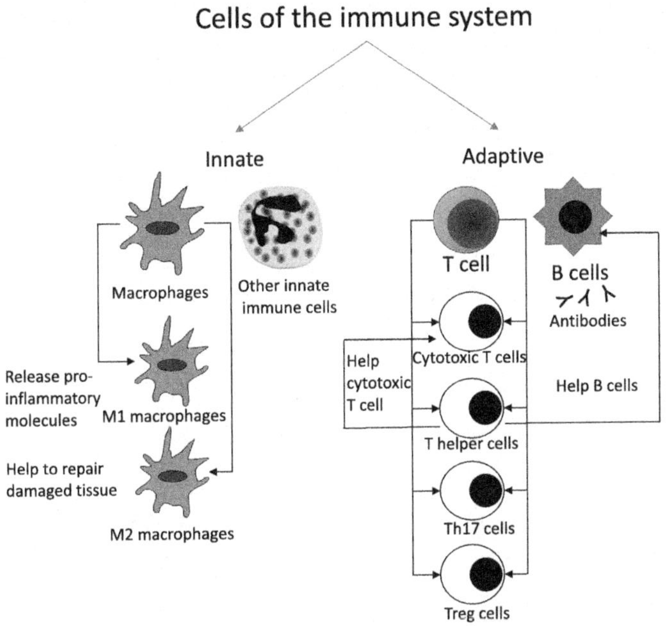

Figure 7.1: Cells of the innate and adaptive immune system. *Innate immune cells are the body's first line of defense. M1 macrophages clean wounds, remove dead cells and release pro-inflammatory molecules. M2 macrophages help to repair damaged tissues. The main cells of the adaptive immune response are the T cells and B cells. B cells produce antibodies. There are a number of different types of T cells. Cytotoxic T cells kill pathogen infected cells and can kill cancer cells, T helper cells help the B cells and the cytotoxic T cells. Th17 are inflammatory T cells, while Tregs down regulate the action of the other T cells. All of these cells are found in the blood and are known as the white blood cells.*

There are a number of different "white" cells (leucocytes) that circulate around our body via the blood which belong to both the innate and adaptive immune systems, see Figure 7.1. One cell type of the innate immune system are macrophages. Macrophages can simplistically be divided into two types: Those associated with inflammation are known as M1 macrophages and are the cells that normally clean a wound of bacteria, debris and dead cells. The second type of macrophages are known as M2, and these are anti-inflammatory and help to renew damaged tissue. It has been established that one way microbiota maintain an environment that is tolerant to the bacteria in our microbiota is by inducing M2 macrophages. Moreover, in mouse models it has been shown that the short chain fatty acids (SCFA) produced by these bacteria trigger the induction of M2 macrophages by directly inducing a chemical from us, their host, via an epigenetic mechanism[5].

One type of cell belonging to the adaptive immune response that circulates in the blood is called the T cell and within these T cells there are different types of cells with different functions. Some, like the cytotoxic T cells, have a role is killing infected cells (often infected with viruses), others such as T-helper cells help the cytotoxic T cells kill and assist in the development of antibody producing B cells. Others, called Th17 cells can cause inflammation, while in contrast, those called Treg cells, do their job of suppressing the immune response thereby maintaining tolerance (see Figure 7.1). The Treg cells have their heyday when we have flooded the circulation with a high level of cytotoxic T cells of the adaptive system which we produced to reject a pathogen. These Tregs put cytotoxic T cells to rest, but these resting T cells act as a reserve, ready to become active again in case the pathogen returns. The immune health of the gut is, in part. maintained by a delicate balance between these types of T cells and the profile of the type of cells induced can be modulated by the microbiota.

In a mouse model it has been shown that short chain fatty acids (SCFA), produced by bacteria in the gut from polysaccharides consumed in our diet (but digested by them) can protect mice from inflammation of the gut by suppressing the production of inflammatory Th17 cells. Furthermore, these SCFA could increase the production of Treg T cells by inducing the expression of a transcription factor characteristic of this type of cell[6]. So, factors produced by the gut microbiome can control by epigenetic means the balance of cells between those that suppress an inflammatory response in the local environment and those that cause inflammation[7]. When this balance is perturbed, conditions such as inflammatory bowel disease can occur resulting in poor and delayed wound healing within the gut.

Inflammatory Bowel Disease - do our Microbiota Play a Role?

The term inflammatory bowel disease (IBD) covers a group of disorders that includes Crohn's disease and ulcerative colitis. It has been called a disorder of how the brain "talks" to the gut as many patients report that psychological stress can trigger the symptoms. These disorders have a complex aetiology, that is there are many factors that contribute to their development including genetic factors, previous gut infections and diet. IBD disorders are characterised by chronic inflammation in the intestines and changes in the structure of chromatin is seen time and again in patients with inflammatory bowel disease. When the DNA methylation is studied in the sera and intestinal cells of patients with IBD, unusual patterns of methylation are seen. In addition, in patients with Crohn's disease, differences in histone methylation have also been seen[8]. But how much of these changes in epigenetic marks is down to the microbiota?

What is observed in patients with IBD is a difference in the type of microbes found in the gut. In one of these studies, biopsies were taken of the inflamed arcas and non-inflamed areas of the colon from patients with Crohn's disease and ulcerative colitis when undergoing colonoscopy. Importantly this was a relatively large study enrolling 50 patients with Crohn's disease, 80 with ulcerative colitis and 31 healthy controls[9]. In this study, some pro-inflammatory bacteria and a decrease in microbiota diversity was identified in the patients compared to the controls. Moreover, within patients, changes in the microbiota between inflamed and non-inflamed areas were observed. Importantly, these changes in microbiota were associated with immune-related epigenetic marks.

The transfer of faecal material is becoming a possible form of treatment for IBD. Indeed, a clinical trial by a Norwegian group with 125 IBD patients enrolled showed that the transplantation of faecal microbiota isolated from healthy donors into the small intestines of IBD patients gave very positive results. This was a placebo-controlled trial, that is some patients did not receive donor microbiome but their own faecal material, and the other 87 patients received donor microbiome. The patients that received donor faecal material had significantly fewer IBD symptoms, showed less fatigue and had a better quality of life compared to the placebo group, three years after the transplant[10]. Moreover, this study identified 10 types of bacterial that were associated with IBD symptoms.

However, working with human subjects it is difficult to analyse the mechanisms involved in the responses seen. Not only for obvious ethical difficulties but also because we are genetically diverse, consume different diets and are exposed to different external stimuli both physically and socially. Such variances can alter our microbiota and alter our susceptibility to disease. Mice can be bred to be genetically very similar, can be housed in the same conditions and exposed to the same diet, removing many of the confounding factors when carrying out experiments on inflammation of the gut. What has been seen in these models is that the Tregs and the Th17 inflammatory cells are plastic and have the ability to change into one another and this is dependent on the presence of a microbiota[11]. In another study, mice were treated with butyrate, a short chain fatty acid (SCFA), and this showed that butyrate could promote the development of T helper cells and suppress the development of inflammatory T cells. Moreover, the scientists showed that butyrate could inhibit the action of T cells involved in the development of colitis. The data showed that butyrate, inhibited the potential of inflammatory Th17 cells to induce colitis. Thus, the content and balance of the microbiota species within the gut is critical to maintaining a healthy gut and certainly may contribute to the development of IBD.

There is therefore convincing data that the microbiome has a role in the development of IBD and studies in mice suggest that this may be via SCFA which we have seen in Chapter 6 can inhibit histone deacetylases thereby increasing gene expression (see Chapter 2).

Microbiota and Cancer – with us or against us?

Cancer has been described as "wounds that do not heal"[12] inferring that the general processes involved in wound healing are similar to that seen in the stroma found in the tumour microenvironment of cancers (see Chapter 4, Figure 4.1). We have seen in the previous section that the microbiota play a significant role in the immune response, so as well as influencing the expression of genes by cancer cells, can it influence the immune response to cancer?

As with the studies investigating the role of the microbiota in the disturbance of the immune system and in the development of inflammatory bowel disease, the investigation of their role in cancer has focused on analysing the link with the gut microbiota (GM). Moreover, of the gastrointestinal and liver cancers most closely linked to the GM, colorectal cancer (CRC) has been the most widely studied. Colorectal cancer is the third most prevalent cancer worldwide in men and the second most common in women (data from 2020). As well as investigating the role of the composition of the gut microbiota in cancer development, its role in determining the response or resistance to cancer therapy of several cancers is also under investigation.

Some bacteria can disrupt the normal symbiosis of gut microbiota with the host

In colorectal cancer it has been possible to ask detailed questions about how the composition of the gut microbiota (GM) associated with cancer (GMCa) differs from the GM associated with the normal gut tissue (GMN) in the same individual. Differences in host (that is human) gene expression in methylation of host DNA, and in the production of short chain fatty acids (SCFAs) by the microbiota, have all been documented and correlated[13]. Analysing the composition of the microbiota showed the presence of a bacterium (Fusobacterium nucleatum) in the GMCa. Interestingly, this bacterium is normally found in the mouth, where it is a part of the oral microbiota, but not elsewhere in the body under healthy conditions. However, in some conditions and in some cancers, for example colorectal and breast cancers, it has been found to migrate from the oral cavity to other tissues, see Figure 7.2. In studies with cell lines derived from colorectal cancer (CRC), this bacterium has been shown to induce cancer cell proliferation, to promote the growth in mice of human CRC samples, and to block the efforts of the immune system to reject pathogens (i.e. block immunosurveillance)[14]. In a mouse tumour model, binding of the bacterium to the tumour has been shown to block tumour infiltrating T cells and promote tumour growth and metastatic progression.

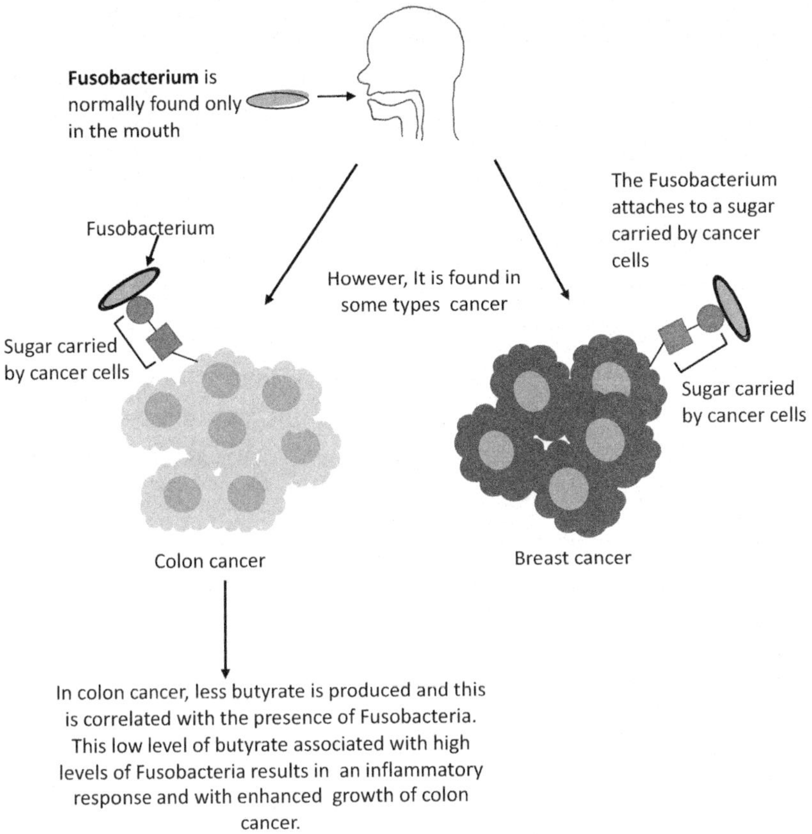

Figure 7.2: Fusobacterium can be found in cancers. *Fusobacterium is normally found in the mouth but also in some cancers including colon and breast. In colon cancer the presence of Fusobacterium is associated with low levels of butyrate.*

In comparing the metabolomic factors produced by tissue associated with GMCa or GMN, there was a marked reduction in the level of butyrate-related colonic metabolites in the GMCa associated tissue which correlated with the abundance of Fusobacterium. Butyrate produced by our good microbiota can be used as a substrate for the production of acetyl CoA by the mitochondria in our healthy gut cells. Within the mitochondria, acetyl CoA is used to produce the energy source, ATP (see Chapter 3, Figure 3.2). Since the cancer cells do not use butyrate as a source of energy, it normally accumulates in the cancer. However, the presence of the Fusobaterium stops this accumulation thus reducing the effects of butyrate on tumour growth which are to reduce inflammation and to induce epigenetic changes in host gene expression which slow the growth of the cancer cells. One obvious question is - does the presence of Fusobacterium affect the response of patients to therapy and in a large clinical trial with over 1000 CRC patients, who were receiving chemotherapy, clinicians have asked this question. The level of Fusobacterium DNA was measured in the GMCa of each patient and compared to their life expectancy. The answer to the question was that higher amounts of Fusobacterium DNA was associated with higher colorectal cancer- specific mortality.

The Fusobacterium has a lectin (a molecule that can bind sugars) called Fap2 which can bind to a sugar expressed on cancers and allows the efficient colonization of the bacterium within the cancer, see Figure 7.2. Since the sugar is expressed on many cancers, particularly adenocarcinomas, it is possible that this bacterium can colonise these cancers[15]. Indeed, it has been found that through its lectin, this bacterium can bind to other cancers expressing the same sugar and gene sequences for Fusobacteria are found in 30% of breast cancers[16].

Cancer treatment based on activating the immune response: The role of gut microbes in response to therapy

We have just discussed how a bacterium in the microbiota of colon cancer can help the cancer cells to proliferate and metastasize, so if an antibiotic were to be developed which specifically killed this bacterium, it might be considered for therapy. This drug would have to be very specific indeed as there are bacteria in the microbiota which are on our side and could help us, rather than our cancer. Such bacteria are being identified in studying how microbiota of different composition can help cancer therapy, and these studies focus on a particular form of therapy.

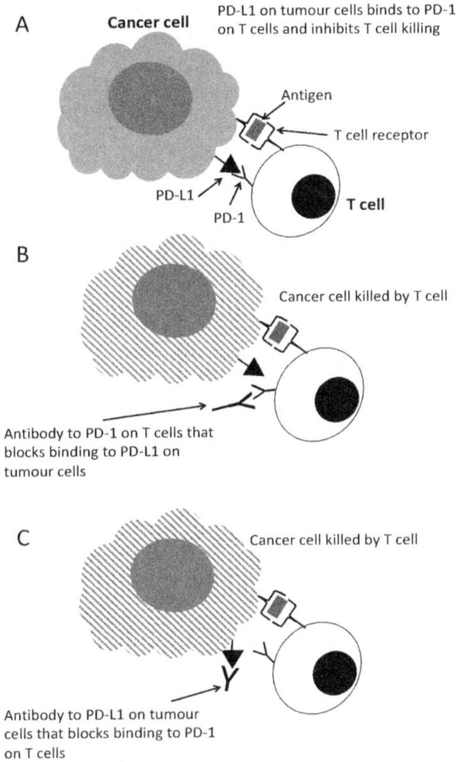

Figure 7.3: Ligands on cancer cells bind to receptors on T cells to inhibit their activation and these interactions are known as immune checkpoints. *A: Cancer cells can express the ligand PD-L1 which when it binds to the PD-1 receptor on T cells causes the T cell to shut down. CD80 on tumour cells and the CTLA4 receptor on T cells work in the same way. B: Therapeutic antibodies to PD-1 on T cells inhibits the binding to PD-L1 on cancer cells and so the T cells become activated and can kill the cancer cells. C: Therapeutic antibodies to PD-L1 on cancer cells inhibit the binding to PD-1 on*

T cells also allowing activation of T cells which can now kill the cancer cells. These antibodies are known as Immune Checkpoint Inhibitors.

As we saw in the section on the immune response, there is a dynamic network in adaptive immunity where cells can activate the immune response and cells like Tregs can dampen this response to maintain a balance to fight infection of pathogens and at the same time be tolerant of self (including the microbiota). However, there is also an additional mechanism to control the T cells. T cells have on their surface molecules called receptors that can bind to other molecules in the circulation or on the surface of other cells. These types of receptors on T cells, (for example those called PD1 and CTLA4,) bind to inhibitory molecules, (called PD-L1 and CD80 respectively, see Figure 7.3).

These inhibitory molecules can be found on other cells and when these bind to the receptor on T cells causes the T cells to shut down, i.e. they can't function. So, the interaction of PD-L1 and CD80 with their receptors on cytotoxic T cells is a "check point" where a balance between activating and curtailing the immune response is normally operative, acting to fight invading pathogens while maintaining tolerance of self. Cancer cells are very clever, and some cancer cells have turned on expression of the inhibitory molecules that bind the receptors found on T cells, see Figure 7.3. Therefore, when the active T cell enters the tumour ready to kill the cancer cells, it encounters the inhibitory molecule on the cancer cell which binds to the T cell receptor causing the T cell to shut down, so it is unable to do its job. Antibodies have been developed by researchers to block the signalling of these two receptors and these are known as Immune Checkpoint Inhibitors (ICI). These ICI are being used as a therapy in a wide range of cancers, including melanoma (a skin cancer), with amazing results and more recently for a type of breast cancer known as triple negative breast cancer.

It has been known for some years that melanoma patients can have a spontaneous partial regression and - very rarely - a full regression of their cancer[17]. Since immune surveillance has been attributed to the regression, attempts have been made down the years to develop therapies based on mobilising the immune response, but these have met with little success[18]. However, with the advent of immune checkpoint inhibitors (ICI), there has been an impressive improvement in the response of some melanoma patients to this type of therapy. However, this response is only seen in a subset of patients.

Exciting results were published in 2015 using mouse models of melanoma that showed that the type of bacteria in the gut influenced the efficacy of this type of treatment[19]. The gut bacteria were influencing how the melanoma responded to the immunotherapy! It sounds quite remarkable but now we know that the microbiome has a strong influence over our immune system. This study has now been replicated many times and in human studies it was found that patients who responded to immune checkpoint therapy had a higher diversity of bacteria in their gut microbiome. Working with patients with metastatic melanoma (given ICI therapy), it has been possible to document in detail the profile of bacteria found in the gut microbiota of responders (R) and non-responders (NR). All of the bacteria have very long names, and we refer you to the references for these names[20,21], but suffice it to say that several bacteria were found to be present in responding (R) patients at higher levels than in the NR patients, and other bacteria were found at higher levels in the NR patients than in the R patients, see Figure 7.4. Considering the bacteria over expressed in R patients as "beneficial" (B) and those increased in NR patients as "non beneficial" (NB), a ratio of numbers of beneficial to non-beneficial (NB) bacteria can be calculated for each patient and any patient with a ratio of B:NB greater than 1.5 was found to be a responder. While this could be useful as a biomarker to predict response to ICI therapy, the complexity of the techniques required and the manpower of scientists with expertise to use them, keeps this within the research community at the moment. As improvement and simplification of techniques move on, it may in the future move into the clinic routinely as a marker.

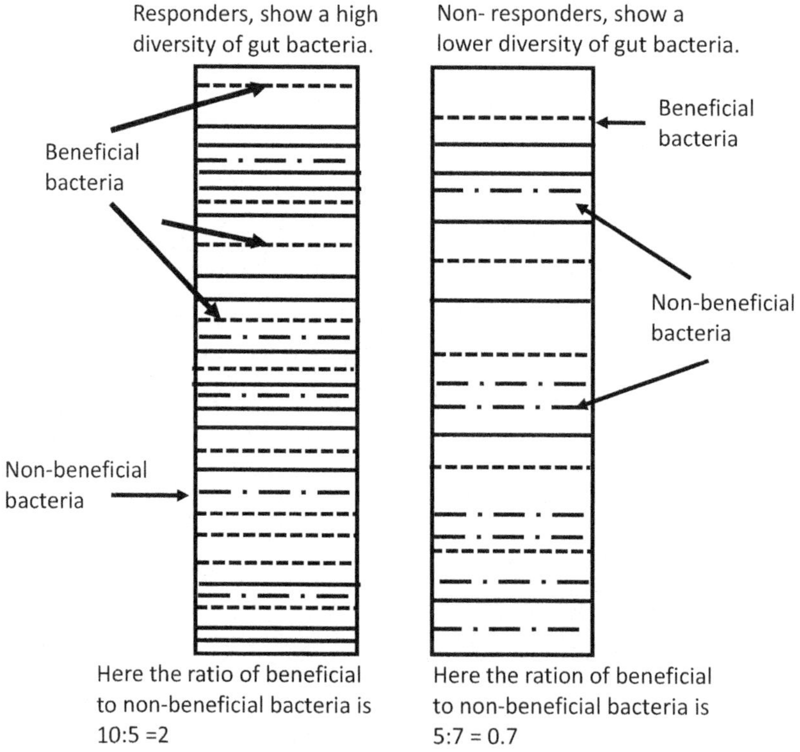

Figure 7.4: Melanoma patients who respond well to immune checkpoint therapy have diverse types of bacteria in their gut microbiota and more beneficial bacteria. *The number of lines in each rectangle represents different types of bacteria. In this diagrammatic representation, in the responders the ratio of*

beneficial bacteria to non-beneficial is 2. Patients with ratios of above 1.5 were shown to be responders.

As with most research with microbiota, the GF mice that have few or no microbiota can provide an experimental model to investigate the influence of different microbiota, and it is possible to transfer the human microbiota from the faeces from R and NR patients to GF mice to check their response to ICI therapy (anti-PDL1 treatment) of melanoma. The results are striking in that ICI treatment of the mice was very effective in mice carrying microbiota from R (responding) patients and completely ineffective in mice with the microbiota from NR (non-responding) patients.

In considering the effect of the microbiome on response to therapy, we can ask "can some of the things we are including in our diet have an effect on response to therapy"? With the availability of commercial pro-biotics i.e. yoghurts containing species of bacteria thought to be beneficial to health, many people, including some cancer patients, are including these in their diet. Another self-help factor which can be included in our diet is fibre (called a pre-biotic) which can feed the microbiota to produce the metabolites we need. Some observational clinical studies have been done to see how addition of either of these dietary factors affect the response of melanoma patients to treatment with immune checkpoint inhibitors (ICI). The two factors have been tested together or singly and the results indicate that the best response (longest time to progression, known as progression free survival, PFS) was seen in patients using fibre in their diet, without taking probiotic yoghurt. Different studies using probiotics alone have been variable, but where it was taken with fibre the probiotics reduced the positive effect seen with fibre alone. It is clearly early days to make any decision regarding whether including probiotics in the diet can affect patient response to therapy, as the bacteria included in the yoghurts vary. However, including fibre in the diet does seem to have a positive effect on PFS in melanoma patients treated with immune checkpoint inhibitors. Fibre will certainly keep the microbiota healthy and productive and could increase the levels of those bacteria which are feeding well thus modifying the composition of the microbiota. Therefore, a mixed diet eating plenty of fibre found in grains, beans, peas and lentils, with plenty of cruciferous vegetables and fruit appears to be the best diet for patients and for the wider population.

That the microbiota play a major role in affecting response to cancer therapy, particularly to therapy aiming to enhance the immune response, is well established. However, the mechanisms involved are still unclear, so it is difficult to say how or if epigenetics play a role but as we saw above the microbiota can influence immune cells via epigenetics, so it is very likely that this does have a role to play. Moreover, as we mentioned earlier in this Chapter, epigenetic change directed by the gut microbiota has been seen, not only in the local host colonic tissue but also in more distant tissues, suggesting some signalling occurring through the circulation[22]. This means that the factors produced by the gut microbiome could have effects on distant tissues where cancers are developing and where therapy is being applied. We know that butyrate produced by the gut microbiome can inhibit the growth of cancers of the colon (see above) and it is possible that circulating butyrate could directly inhibit the growth of melanoma cells. No doubt each tissue will have its own specific reaction to the signals coming from the microbiota.

To investigate the epigenetic mechanisms involved in the influence of the microbiota on the outcome of patients treated with immune checkpoint inhibitors, large numbers of patients are required to produced definitive results. The easiest way to look for an effect of the gut microbiota (GM) is to look for epigenetic changes in the blood of patients and compare this to their microbiome. The differential methylation of host DNA or the profile of microRNA (miRNA) in the circulation could be correlated with the bacterial profile of the microbiota. When looking at the effect of the microbiota on the response to check-point immunotherapy, nearly all experiments and studies have been performed looking at the gut microbiome. We wonder if there is any value in looking at the skin microbiome in response to check-point therapy in melanoma patients?

There are many bacteria and standardizing what constitutes beneficial and nonbeneficial bacteria could be a useful step forward in knowing how to use this information for improving cancer therapy. However, although the link between the microbiome and response to therapy is well established there is still much work to be done to define the mechanisms involved.

Glossary for Chapter 7

Acetyl CoA	A precursor for the TCA cycle where it is oxidised for energy production.
Adaptive immunity	Cells of the immune system that fight infection. These cells have memory.
Aetiology	The cause or causes of a certain disease.
ATP	The molecule in which the body's energy is stored.
B-cells	These are the cells of the adaptive immune response that produce antibodies.
Butyrate	A short chain fatty acid produced by the fermentation carried out by bacteria in the microbiota.
CD80	The molecule or ligand that can bind to CTLA4 on T cells.
Chromatin	The basic unit of the nuclear genome containing DNA and histone proteins.
CTLA-4	A receptor on the surface of T cells that when activated by its binding partner can shut down the T cell.
Cytotoxic T cells	Cells of the adaptive immune response and the role of these cells is to kill infected cells.
Enzyme	A substance (usually a protein) which catalyses a specific biochemical reaction so that the product is produced very quickly, but the enzyme is not changed in anyway.
Fusobacterium	A bacterium normally found in the mouth but has been found in some cancers e.g. colon, breast.
Germ-free mice	Mice that have been born and raised in germ-free conditions. They have no microbiota.

Helper T cells	Cells of the adaptive immune response that help the cytotoxic T cells and B cells to mature.
Immune surveillance	How the host's immune system looks for and recognises foreign pathogens.
Immune system	A complex system of cells and tissues that help the body fight infections and other diseases.
Inflammatory bowel disease	A term for two conditions, Crohn's disease and ulcerative colitis, that cause chronic inflammation of the gastrointestinal track.
Innate immunity	Cells and substances that are the body's first line of defence against infection.
Lectin	A molecule, usually a protein that can bind to a specific sugar or sugars.
M1 macrophages	These cells of the innate immune system are associated with inflammation and release pro-inflammatory substances.
M2 macrophages	These cells of the innate immune system are anti-inflammatory and help to renew damaged tissue.
Macrophages	A cell type of the innate immune response.
Microbiota	The bacteria which live with us in symbiosis in the gut and other tissues.
Mucus	The sticky substance that lines the gut and many other tissues that have "outside" edges.
Pathogen-germ mice	Mice that have no pathogenic organisms but do have a healthy microbiota.
PD-L1	The molecule or ligand that can bind to PD1 on T cells, often produced by cancer cells.

PD1	A receptor on the surface of T cells that when activated by its binding partner can shut down the T cell.
Placebo	A treatment that appears real to the patient but is designed to have no therapeutic benefit.
Polysaccharides	A carbohydrate which consists of a number of sugar molecules (eg starch or cellulose) bound together.
Short chain fatty acids (SCFA)	SCFA are a subset of fatty acids produce by the microbiota by the digestion of food such as polysaccharides and fibre that we cannot digest.
T cells	Cells of the adaptive immune response.
Th17 cells	These are T cells that are pro-inflammatory and secrete pro-inflammatory substances.
Transcription factor	A substance, usually a protein, which can bind directly to chromatin to regulate gene expression.
Tregs	These are T cells that can suppress the immune response thereby maintaining self-tolerance.
"White" cells circulating in the blood	`These are "leucocytes" from the Greek Leuco/white and cytera /cells.

Epilogue

"Try to know something about everything and everything about something".

T H Huxley

Now you have come to the end of this book we hope you have enjoyed getting to know a bit more than "something" about Epigenetics. More and more of us are wanting to know about science, not just biology but also astronomy, physics and chemistry. As Edwin Hubble of telescope fame said:

"We do not know why we are born into the world, but we can try to find out what sort of a world it is"

We hope we have conveyed a little about the complex way genes are controlled and how each tissue and even each cell can express different sets of genes. Without Epigenetics life could not exist beyond the single cell state.

In considering all the chapters in this book there are some features which appear to run through them all. In looking at the time scale of the important discoveries and their impact on advances in knowledge, one feature emerges throughout and that is how developments in technology influence progress in our understanding of biology and indeed all sciences. Just as sequencing DNA has opened doors in genetics, so, in Epigenetics, sequencing DNA to identify methylated cytosines has opened doors for defining the role of this modification in controlling gene expression in normal tissue development and in disease. Perhaps less flamboyant, but with enormous impact, the improvements in culturing cells *in vitro* allowed experimental studies to be undertaken, initially for investigating viruses, but also for asking multiple questions about how cells work. With the availability of cell culture, tumour suppressor genes and oncogenes were identified. Now, in using cell culture to study cancer, methods are being developed for culturing not only the cancer cells, but organoids which include the other cells in the tumour with which the cancer cells interact and influence.

It is notable that there is often a considerable gap in time between the years when individuals started their seminal work and the time when their work was acknowledged by awarding the Nobel prize. As time goes on and new technologies develop, the importance of their original work expands and the 2023 Nobel prize for Medicine is a fine example of this, where the innovative work on mRNA was started over 30 years ago culminating in the development of the Covid-19 vaccines that saved many of us. As we write this book, the first cancer patients have been administered a personalised mRNA cancer vaccine as part of a clinical trial. It will take a number of years and trials to determine if this is a breakthrough in cancer therapy, but we live in hope. Of course, lots of people have done important work which has not won them a Nobel prize, but that doesn't mean that their work was not seminal. What about Avery who showed that the important genetic material was DNA? He was not awarded the Nobel prize but is recognised by scientists who acknowledge his important contribution.

Another feature which we should recognise is the enormous contribution of our fellow creatures to the advancement of the study of Epigenetics. Apart from mice which, being mammals, have been widely used for investigating what happens with genes *in vivo*, much of the early work in Epigenetics, particularly in embryological development was done with toads (who have very large eggs). Following this early work, the worm has yielded amazing findings in giving us microRNA and a novel method of controlling the expression of proteins by inhibiting the function of their coding mRNA. Work with the fruit fly, Drosophila, opened the doors to identifying epigenetic factors controlling gene expression. It is interesting to find that many of the long non-coding RNAs are only found in primates, so finding out how they function *in vivo* is a challenge. Nonetheless this is an important finding, suggesting that these genes which are new in the evolutionary scale may be important for brain development.

It might have come of a surprise to some readers to learn that we have another source of DNA in our cells apart from the DNA in chromosomes, namely the DNA in our mitochondria and that this was brought in by bacteria - offering their incredibly efficient system of oxidative phosphorylation for producing energy from our food. Another surprise could be that all of us, boys and girls alike, inherit our mitochondrial DNA only from our mother. On the other hand, some readers might have already had their mitochondrial DNA sequenced and know which tribe of Homo sapiens they belong to. In Chapter 3 we referred to the tribes of the Seven Daughters of Eve in Europe but sequencing mitochondrial DNA globally has classified further groups of Homo sapiens as kindred through mutual variations in the sequence of their mtDNA. The interaction between the chromosomal, nuclear DNA and the mitochondrial DNA brings in a different perspective on how epigenetic mechanisms are involved in their combined control of gene expression. Considering that the fundamental event of endosymbiosis happened for multi-cellular organisms 1.5 billion years ago, one wonders how many other life changing events happened that we don't yet know about. We suspect that the bacteria living alongside us in our gut in symbiosis have been with us for a long time.

Of the many factors that contribute to the epigenetic control of our genes, there are some that we can influence. Diet is one of these factors and is extremely important for maintaining a healthy body. Although we can have control of diet most of the time, in times of famine, caused by war or natural events, we can only observe - control is out of our hands. Nevertheless, as described in Chapter 5, there is some evidence that children and grandchildren can be influenced by the famine endured by their grandparents and in the case of grandsons from grandfathers, this can be surprisingly good news.

The emphasis on diet is precisely because it is possible to control what we eat but there is no denying that food supplements can make an important contribution to our diet. For example, it is highly recommended that vitamin D supplements are taken in many countries. Vitamin D deficiency is common in many parts of the world although the involvement of vitamin D in the analysis of the epigenetic control of our genes is still in its infancy. It was certainly a deficiency in people who worked in the mills and mines in the industrial revolution in the UK, who showed the signs of this deficiency by being bowlegged, as they saw little sunlight because of their working hours. However, there is evidence that some nutrients, often given as supplements, such as folic acid, vitamin B and green tea have a direct effect on our epigenome. Thus, we have tried to emphasise in many parts of this book the importance of a healthy diet, and this drew us into considering our microbiota, that also depend on what we eat. We have seen that the microbiome can influence the epigenome and that the composition of the microbiome is crucial for human health. The transplantation of the gut microbiota from healthy people to those with a particular disease is becoming available and with the further understanding of the complex ecosystem that is the gut microbiota, the real possibility of pro-biotics for the treatment of some diseases such as IBS is not a too distant dream.

A large proportion of the studies in Biomedicine are, naturally, focused on disease, but in comparing various parameters found in normal and diseased tissue, much is learned about basic underlying mechanisms. The whole fabric of the studies leading to the identification of oncogenes and tumour suppressor genes came from working with tumour viruses and led to the identification of proto-oncogenes, i.e. normal genes which were mutated to produce oncogenes which are involved in cancer growth. Some of these proto-oncogenes are in control of the normal cell cycle or like p53 are "guardians of the genome", making sure that DNA breaks are repaired. In this book we have devoted a Chapter to reporting on the development of cancer which can arise in so many tissues. It is clear that the epigenetic control of genes can contribute to the progression of cancer. Tumour suppressor genes can be switched off and oncogenes over expressed, and therapeutic drugs to target such changes are already being used in cancer therapy. As more is learned about the mechanisms involved and specific drugs are developed, we would predict that the epigenetic control of genes will play an increasing role in the development of new drugs for cancer therapy.

So, what is the future of epigenetics and where will the research take us in the future? Of course, it is impossible to know the future and prophecics are made in the current environment without being able to consider how future developments will influence directions. The vastly expanding areas of technology that are making it faster to perform and analyse experiments will take the advancement of knowledge of Epigenetics into areas it is difficult to predict. Who would have known that in 2003, when the first human genome was finally published after 13years of sequencing, that now a human genome can be sequenced in a day! Of course, in analysis of ancestral DNA – the companies only sequence bits and it takes even less time!!

However, as with any science, when we look forward and predict a revolution that may be helpful to an individual we must proceed with care. There is accumulating evidence, especially in animal models, that the epigenetic profile, modified in an individual, can be inherited in a transgenerational manner. If our advancing knowledge of epigenetic drugs leads to the correction of an epigenetic mistake, will we, as a society, be happy that this correction may be inherited? The immediate correction may be very advantageous for the individual, but should we worry about the potential for altering expression of genes affecting other characteristics and its impact on future generations? This of course is something that society needs to debate, not just in the area of epigenetic control but also in the possibility of future gene "editing". Nevertheless, further work in Epigenetics can only enhance our sense of the wonder of life and its mysteries and with reverence we can still:

"Try to find out what sort of a world it is we live in"

Reference and notes

Chapter 1

1. Vitezslav O. 1984, Mendel. Oxford University Press.
2. Avery OT. et al 1944, Studies on the chemical nature of the substance inducing transformation of pneumococcal types: Induction of transformation by a desoxyribnucleic acid fraction isolated from pneumococcus type III. J Exp Med 79, 137-158.
3. Watson JD. and Crick FHC. 1953, Molecular structure of nucleic acids: Structure for deoxyribose nucleic acid. Nature 171, 737-738.
4. https://wellcomecollection.org/works/v3ndr2ux
5. Cobb M. and Comfort N. 2023, What Rosalind Franklin truly contributed to the discovery of DNA's structure. Nature 616, 657-660.
6. Crick FHC et al 1961, General nature of the genetic code for proteins. Nature 192,1227–1232
7. https://www.nobelprize.org/prizes/medicine/1968/nirenberg/facts/
8. https://www.nobelprize.org/prizes/medicine/2023/press-release/
9. Olins AL. and Olins DE. 1974, Spheroid chromatin units (v-bodies). Science 183, 330-332.
10. https://le.ac.uk/dna-fingerprinting/history
11. Hall JM. et al 1990, Linkage of early-onset familial breast cancer to chromosome 17q21. Science 250, 1684-1689.

12. Convention dictates that the names of human genes are always written in upper case and italics. Human proteins are in upper case but not italics.
13. Easton DF. et al 1993, Genetic linkage analysis in familial breast cancer and ovarian cancer: Results from 214 families. The breast cancer linkage consortium. Am J. Human Genet. 52, 678-701.
14. Lopez A.F.C. 2020, Mitochondrial metabolism and DNA methylation: a review of the interaction between two genomes. Clin Epigenetics 12:182
15. Endosymbiosis two organisms living together with one inside the other, in this case the bacteria living within the cell.
16. Taanman J-W. 1999, The mitochondrial genome: structure, transcription, translation and replication. Biochimica et Biophysica Acta – Bioenergentics 1410, 103-123.
17. Bryan Sykes, The Seven Daughters of Eve. Bantam Press.

Chapter 2

1. Avery OT. et al 1944, Studies on the chemical nature of the substance inducing transformation of pneumococcal types: Induction of transformation by a desoxyribnucleic acid fraction isolated from pneumococcus type III. J Exp Med 79, 137-158.
2. Robertson R. 1977, Biography of Conrad Hal Waddington.
https://royalsocietypublishing.org/doi/pdf/10.1098/rsbm.1977.0022
3. Waddington CH.1942, The Epigenotype. Inter J Epidemiology 2012, 41, 10–13. The Reprint of the1942 paper.

4. Waddington CH. 1940, Organisers & Genes. Cambridge University Press
5. https://www.nobelprize.org/prizes/medicine/1933/morgan/article/
6. Morgan TH. 1934, Embryology and Genetics. Columbia University Press
7. Waddington CH. 1957, The strategy of the genes: a discussion of some aspects of theoretical biology. Publishers George Allen and Unwin. Wellcome Collection.
8. Bird AP and Southern EM. 1978, Use of restriction enzymes to study enkaryotic DNA methylation: I. The methylation pattern in ribosomal DNA from Xenopis Laecis. J Mol Biol 118, 27-47.
9. Gritschier J. 2009, On the track of DNA methylation. An Interview with Adrian Bird. PLOS Genetics, https://doi.org/10.1371/journal.pgen.1000667
10. Neul JL. 2012, The relationship of Rett syndrome and MECP2 disorders to autism. Dialogues Clin Neurosci.14, 253–262.
11. Vashi N. and Justice MJ. 2019, Treating Rett syndrome: from mouse models to human therapies. Mammalian Genome, 30:90–110
12. Allfrey VG. et al. 1964, Acetylation and Methylation of Histones and their possible role in the regulation of RNA synthesis. Biochemistry 51, 786-794.
13. Anderson JD et al. 2001, Effects of histone acetylation on the equilibrium accessibility of nucleosomal DNA target sites. J Mol Biol 307, 977-85.
14. Schuettengruber B. et al. 2017, Regulation by Polycomb and Trithorax: 70 Years and Counting. Cell 171, 34-57.

15. Taylor-Papadimitriou J. and Burchell JM. 2022, Histone Methylases and Demethylases Regulating Antagonistic Methyl Marks: Changes Occurring in Cancer. Cells 11,1113.
16. Djebali S.et al 2012, Landscape of transcription in human cells. Nature 6, 489, 101-108.
17. Crick FH. et al. 1961, General *nature* of the genetic code for proteins. *Nature. 192, 1227*-32.
18. Lee RC. et al.1993, The C. elegans heterochronic gene lin-4 encodes small RNAs with antisense complementarity to lin-14. Cell 1993, 75, 843-54.
19. Pasquinelli AE. et al. 2000, Conservation across animal phylogeny of the sequence and temporal regulation of the 21 nucleotide let-7 heterochronic regulatory RNA. Nature 408, 86-89.
20. Loda A. and Heard E. 2019, Xist RNA in action: Past, present, and future. PLoS Genet. 15, e1008333.
21. Vallianatos CN. et al. 2018, Altered gene-regulatory function of KDM5C by a novel mutation associated with autism and intellectual disability. Front Mol Neurosci.11, 104.
22. Vrba L. et al. 2015, A lncRNA connected to cell mortality and epigenetically silenced in most common human cancers. Epigenetics 10, 1074-1083.
23. Derrien T. et al. 2012, The GENCODE v7 catalog of human long noncoding RNAs: ananlysis of their gene structure, evolution and expression. Genome Res, 22, 1775-1789.
24. Quinodoz SA. et al. 2021, RNA promotes the formation of spatial compartments in the nucleus. Cell 184, 5775–5790.

Chapter 3

1. Sagan L. 1967, On the origin of mitosing cells. Journal of Theoretical Biology 14, 255–274.
2. https://www.ncbi.nlm.nih.gov/books/NBK9841
3. Martin W. and Mentel M. 2010, The Origin of Mitochondria. Nature Education 3(9):58.
4. Martin W and Herrmann R.G. 1998, Gene Transfer from Organelles to the Nucleus: How Much, What Happens, and Why? Plant Physiol.118, 9–17.
5. Gray MW. et al Mitochondria evolution. Science, 283,1476–1481.
6. Jeon KW. and Lorch IJ. 1967, Unusual intra-cellular bacterial infection in large, free-living amoebae. Experimental Cell Research 48, 236-40.
7. Taanman J.W. 1999, The mitochondria genome: structure, transcription, translation and replication. Biochimica et Biophysica Acta 1410, 103-123.
8. Yamauchi A. 2005, Rate of gene transfer from mitochondria to nucleus: effects of cytoplasmic inheritance system and intensity of intracellular competition. Genetics, 171,1387-96.
9. D'Souza A.R. and Minczuk M. 2018, Mitochondrial transcription and translation: Overview Essays. Biochem 62, 309–320.
10. Coppedê F. 2024, Mitochondria DNA methylation and mitochondrial-related epigenetics in neurodegeneration. Neural Regenerative Research 19:405-406.
11. Yue et al. 2022, Mitochondrial genome undergoes de novo DNA methylation that protects mtDNA against oxidative damage during the peri-implantation window, Proc Natl Acad Science, 119, e2201168119.

12. Tomar A. et al. 2024, Epigenetic inheritance of diet-induced and sperm-borne mitochondrial RNAs. Nature https://doi.org/10.1038/s41586-024-07472-3.
13. Bryan Sykes, The Seven Daughters of Eve. Bantam Press.
14. Lee WT. et al. 2017, Mitochondrial DNA haplotypes induce differential patterns of DNA methylation that result in differential chromosomal gene expression patterns. Cell Death Discovery 3,17062, doi:10.1038.
15. Gill et al. 1994, Identification of the remains of the Romanov family by DNA analysis. Nature Genetics 6, 130-135 (https://blog.sciencemuseum.org.uk/the-forensic-investigation-into-the-romanov-mystery)
16. Hagelberg E. et al. 1989, Ancient bone DNA amplified. Nature 342, 465.
17. Fetterman JL. and Ballinger SW. 2019, Mitochondrial genetics regulate nuclear gene expression. Proc Natl Acad Science, 116, 15765-15765.
18. Borcherding N. and Brestoff JR. 2023, The power and potential of mitochondria transfer. Nature 623, 283-291.
19. Zhang Q. et al. 2010, Circulating mitochondrial DAMPS cause inflammatory responses to injury. Nature 464, 104-108.
20. Tinker RJ. et al. 2021, Current and Emerging Clinical Treatment in Mitochondrial Disease. Molecular Diagnosis & Therapy, 25,181–206.
21. Picard M.et al. 2014, Progressive increase in mtDNA 3243A>G heteroplasmy causes abrupt transcriptional reprogramming. Proc Natl Acad Science, 111, E4033-E4042.

22. Kopinski PK. et al. 2019, Regulation of the nuclear epigenome by mitochondrial DNA heteroplasmy. Proc Natl Acad Science, 116, 16028-16035.
23. Gurdon J. 2012 Transplantation of Nuclei and cell Differentiation. Scientific America Oct 9th.
24. Craven L.et a.l 2016, Research into Policy; A Brief History of Mitochondrial Donation. Stem Cells, 265–267.
25. Costa-Borges N. et al. 2023, First pilot study of maternal spindle transfer for the treatment of repeated *in vitro* fertilization failures in couples with idiopathic infertility. Fertility and Sterility, 119, 964-973.

Chapter 4

1. American Cancer Society. Cancer.org 1 1.800.227.2345.
2. Parreno V. et al. 2024, Transient loss of Polycomb components induces an epigenetic cancer fate. Nature, https://doi.org/10.1038/s41586-024-07328-w
3. Dawson M. et al. 2011, Inhibition of BET recruitment to chromatin as an effective treatment for MLL-fusion leukemias. Nature, 478, 529-33.
4. Dawson M and Kouzarides T. 2012, Cancer Epigenetics: from Mechanism to Therapy. Cell 150, 12-27.
5. Ramaiah MJ.at al. 2021, Epigenetic modulation and understanding of HDAC inhibitors in cancer therapy. Life Sciences 277, 119504.
6. Lane D P. and Crawford L V. 1979, T antigen is bound to a host protein in SV40-transformed Cells. Nature, 278, 261-263.
7. Crawford L.V. et al 1981, Detection of a common feature in several human tumor cell lines a 53,000-daltonprotein Proc. Natl. Acad. Sc. USA, 78, 41-45.

8. Calloway E. 2015, How Elephants avoid Cancer. Nature News, 08 October 2015.
9. Brubaker J. 2019, The 7 Viruses That Cause Human Cancers. American Society for Microbiology Jan 25, 2019.
10. Chatterjee K.et al. 2019, The interplay between Epstein-Bar virus (EBV) with the p53 and its homologs during EBV associated malignancies. Heliyon 5, e02624.
11. Falcaro M. et al. 2021, The effects of the national HPV vaccination programme in England, UK, on cervical cancer and grade 3 cervical intraepithelial neoplasia incidence: a register-based observational study. The Lancet, 398, 2084-2092.
12. Pal A. and Kundu R.2020, Human Papillomavirus E6 and E7: The Cervical Cancer Hallmarks and Targets for Therapy. Frontiers in Microbiology,10, 3116.
13. Maroni R. et al. 2021, A case-controlled study to evaluate the impact of the breast screening programme on mortality in England. Brit. J Cancer, 124, 736–743.
14. Pastorino U. 2022, Baseline computed tomography screening and blood microRNA predict lung cancer risk and define adequate intervals in the BioMILD trial. Annals of Oncology, 33(4):395-405.
15. ClinicalTrials.gov Identifier: NCT02247453
16. McCartney M. and Cohen D. 2024, Galleri promises to detect multiple cancers-but new evidence casts doubt on this much hyped blood test. BMJ, 4,386:q1706.
17. Cheng Y. et al. 2019, Targeting epigenetic regulators for cancer therapy: Mechanisms and advances in clinical trials. Signal Transduct. Target. Ther, 4, 62.

18. Zhang Z. et al. 2022, Recent progress in DNA methyltransferase inhibitors as anticancer agents. Front Pharmacol. 13, 1072651.
19. Eckschlager T. 2017, Histone Deacetylase Inhibitors as Anticancer Drugs. *Int. J. Mol. Sci.*, *18*, 1414.
20. Bonekamp NA. et al. 2020, Small-molecule Inhibitors of Human Mitochondrial DNA Transcription. Nature 588, 712–716.
21. Daglish S C. D. et al. 2023, Targeting Mitochondrial DNA Transcription by POLRMT Inhibition or Depletion as a Potential Strategy for Cancer Treatment. Biomedicines, 11, 1598.
22. Lee et al. 2015, Mitochondrial DNA copy number is regulated by DNA methylation and demethylation of POLGA/POLRMT in stem and cancer cells and their differentiated progeny. Cell Death Dis.,6, e1664.
23. Yu et al. 2021, Targeting mitochondrial function along with other therapies can overcome resistance. Front Chem. 9, 775226.
24. Jin P. et al. 2022, Mitochondrial adaptation in cancer drug resistance: prevalence, mechanisms and management. Journal of Hematology & Oncology, 15, 97.
25. Gupta RA. et al. 2010, Long non-coding RNA HOTAIR reprograms chromatin state to promote cancer metastasis. Nature, 464, 1071-6.
26. Pastori C. et al. 2015, The Bromodomain protein BRD4 controls HOTAIR, a long noncoding RNA essential for glioblastoma proliferation. Proc. Natl. Acad. Sci. USA, 112, 8326-833.
27. Zhao J. et al. 2021, Combination LSD1 and HOTAIR-EZH2 inhibition disrupts cell cycle processes and induces

apoptosis in glioblastoma cells. Pharmacol. Res, 171,105764.

28. https://www.mskcc.org/news/sweet-revenge-taking-advantage-cancer-cells-hunger-sugar-and-other-nutrients
29. Li Y. Liu L V. and Tollesfsbol O. 2010, Glucose restriction can extend normal cell lifespan and impair precancerous cell growth through epigenetic control of hTERT and p16 expression. FASEB J. 24, 1442-53.
30. https://news.cancerresearchuk.org/2023/01/30/starving-cancer-into-submission/

Chapter 5

1. Landro L. 2010, "Healthy Living, for Two." Wall Street Journal, September 9th
2. De Rooji et al. 2022, Lessons learnt from 25years of research into long-term consequences of prenatal exposure to the Dutch Famine 1944-1945: The Dutch famine birth cohort. Int J of Environ Health Res 32,7, 1432-1446.
3. Tobi EW.et al. 2014, DNA methylations signatures link prenatal famine exposure to growth and metabolism. Nature Commun. 5:5592 | DOI: 10.1038/ncomms6592
4. Heijmans BJ. et al. 2008, Persistent epigenetic differences associated with prenatal exposure to famine in humans. Proc Natl Acad Sci 105, 17046-17049.
5. Bygren LO. et al. 2001, Longevity determined by paternal ancestors' nutrition during their slow growth period. Acta Biotheoretica 49, 53-59.
6. Kaati G. et al. 2002, Cardiovascular and diabetes mortality determined by nutrition during parents' and grandparents' slow growth period. Euro J Hum Genetics 10, 682-688.

7. Vagero D. et al 2018, Paternal grandfathers' access to food predicts all-cause and cancer mortality in grandsons. Nature Commun DOI: 10.1038/s41467-018-07617-9
8. Eckersley-Laslin MA. et al. 2018, Dynamics of the epigenetic landscape during the maternal-to-zygote transition. Nature Reviews Mol Cell Biol 19, 436-450.
9. Al-Mousawi J. and Boskovic A. 2022, Transcriptional and epigenetic control of early life cell fate decisions. Current Opinions Oncol 34,148-154.
10. Ishida M. and Moore GE. 2013, The role of imprinted genes in humans. Mol Aspects of Med 34, 826-840.
11. El Khattabi LA. et al. 2019, A genome wide search for new imprinted genes in the human placenta identifies DSCAM as the first imprinted gene on chromosome 21. Euro J Human Genetic 27,49-60.
12. Zeng Y. and Chen T. 2019, DNA methylation reprogramming during mammalian development. Genes 10, 275.
13. Fitz-James MH. And Cavalli G. 2022, Molecular mechanisms of transgenerational epigenetic inheritance. Nature Reviews Genetics 23,325-341.
14. Senaldi L. and Smith-Raska M. 2020, Evidence for germline non-genetic inheritance of human phenotypes and diseases. Clin Epigenetics 12,136.
15. Dias BG. and Ressler KJ. 2014, Parental olfactory experience influences behaviour and neural structure in subsequent generations. Nature Neuroscience 17, doi:10.1038/nn.3594.
16. Tomar A. et al. 2024, Epigenetic inheritance of diet-induced and sperm-borne mitochondrial RNAs. Nature https://doi.org/10.1038/s41586-024-07472-3.

17. https://www.who.int/news-room/fact-sheets/detail/obesity-and-overweight

Chapter 6

1. Cheung P.et al. 2018, Single-Cell Chromatin Modification Profiling Reveals Increased Epigenetic Variations with Aging. Cell 73, 1385–1397.
2. Alegría-Torres A.J. 2011, Epigenetics and lifestyle. Epigenomics 3, 267-77.
3. Dashwood R.H. and Ho E. 2007, Dietary histone deacetylase inhibitors: from cells to mice to man. Seminars in Cancer Biology, 17, 363-369.
4. Gibbs A. et al. 2009, Sulforaphane destabilizes the androgen receptor in prostate cancer cells by inactivating histone deacetylase Proc Natl Acad Sci U S A. 106, 16663–16668.
5. Pandey M. et al. 2010, Promoter demethylation and chromatin remodelling by green tea polyphenols leads to re-expression of GSTP1 in human prostate cancer cells. Int J Cancer, 126, 2520-2533.
6. Pascale A. et al. 2018, Microbiota and Metabolic diseases. Endocrine, 61, 375-371.
7. Gill S.R. et al. 2006, Metagenomic Analysis of the Human Distal Gut Microbiome. Science, 312, 1355–1359.
8. Forster SC. et al. 2019, A human gut bacterial genome and culture collection for improved metagenomic analyses. Nature Biotech. 37, 186–192.
9. Oliphant,K. and Allen-Vercoe E. 2019, Macronutrient metabolism by the human gut microbiome: major fermentation by-products and their impact on host health. Microbiome, 7, 91.

10. Krautkramer KA. et al. 2016, Diet-Microbiota Interactions Mediate Global Epigenetic Programming in Multiple Host Tissues. Mol Cell. 64, 982-992.

11. Junca H. et al. 2022, The emerging potential of microbiome transplantations. Comput Struct Biotechnol J. 20, 615–627.

12. Cammarota G. et al. 2015, Randomised clinical trial: faecal microbiota transplantation by colonoscopy vs. vancomycin for the treatment of recurrent Clostridium difficile infection. Aliment Pharmacol Ther. 41, 835-43.

13. Zhang F. et al. 2012 Should we standardize the 1,700-year-old fecal microbiota transplantation? American Journal of Gastroenterology. 107, 1755,

14. Jones JM. 2014, CODEX-aligned dietary fiber definitions help to bridge the 'fiber gap'. Nutr J. 13, 34.

Chapter 7

1. Zitvogel L. and, Kroemer G. 2021, Commensals shape the immune system. Nat Rev Immunol 21, 615.

2. Round J. and Mazmanian S. 2009, The gut microbiota shapes intestinal immune responses during health and disease. Nat Rev Immunol 9, 313–323.

3. Ansari I. et al. 2020, The microbiota programs DNA methylation to control intestinal homeostasis and inflammation. Nat Microbiol. 5, 610-619.

4. lajnik MF. and Kasahara M. 2010, Origin and evolution of the adaptive immune system: genetic events and selective pressures. Nat Rev Genet.11,47-59.

5. Amamou A. et al. 2022, Gut Microbiota, Macrophages and Diet: An Intriguing New Triangle in Intestinal Fibrosis. Microorganisms 10, 490.

6. Round JL. and Mazmanian SK. 2010, Inducible Foxp3+ regulatory T-cell development by a commensal bacterium of the intestinal microbiota. Proc Natl Acad Sci U S A. 107, 12204-9.
7. Guo T. et al. 2023, Dietary polysaccharides exert biological functions via epigenetic regulations: Advance and prospectives. Crit Rev Food Sci Nutr. 63, 114-124.
8. Amatullah H. and Jeffrey KL. 2020, Epigenome-metabolome-microbiome axis in health and IBD. Current Opinion in Microbiology, 56,97-108.
9. Ryan FJ. et al. 2020, Colonic microbiota is associated with inflammation and host epigenomic alterations in inflammatory bowel disease. Nat Commun. **11**, 1512.
10. El-Salhy M. et al. 2022, Efficacy of Fecal Microbiota Transplantation for Patients with Irritable Bowel Syndrome at 3 Years After Transplantation. Gastroenterology. 163, 982-994.
11. Ueno A. et al. 2018. TH17 plasticity and its relevance to inflammatory bowel disease. J of Autoimmunity, 87, 38-49.
12. Dvorak HF.1986, Tumors: wounds that do not heal. Similarities between tumor stroma generation and wound healing. N Engl J Med. 315, 1650-9.
13. Wang Q. et al. 2020, Multi-omic profiling reveals associations between the gut mucosal microbiome, the metabolome, and host DNA methylation associated gene expression in patients with colorectal cancer. BMC Microbiol 20, (Suppl 1), 83.
14. Kim HS. et al. 2023, Fusobacterium nucleatum induces a tumor microenvironment with diminished adaptive immunity. Front Cell Infect Microbiol. 13, 1101291.

15. Abed J. et al. 2017, Tumor Targeting by Fusobacterium nucleatum: A Pilot Study and Future Perspectives. Front Cell Infect Microbiol. 7, 295.
16. Parhi L. et al. 2020, Breast cancer colonization by Fusobacterium nucleatum accelerates tumor growth and metastatic progression. Nature Commun. 11, 3259.
17. Sumner WC.1953, Spontaneous regression of melanoma. Report of a case. Cancer 6, 1040-1043.
18. Byrne EH. and Fisher DE. 2017, Immune and molecular correlates in melanoma treated with immune checkpoint blockade. Cancer 123(S11):2143-2153.
19. Vétizou M. et al. 2015, Anticancer immunotherapy by CTLA-4 blockade relies on the gut microbiota. Science 350 (6264),1079-84.
20. Matson V.et al, 2018, The commensal microbiome is associated with anti–PD-1 efficacy in metastatic melanoma patients. Science 359 (6371),104-108.
21. Gopalakrishnan V. et al. 2018, Gut microbiome modulates response to anti–PD-1 immunotherapy in melanoma patients Science, 359(6371), 97-103).
22. Krautkramer KA. et al. 2016, Diet-Microbiota Interactions Mediate Global Epigenetic Programming in Multiple Host Tissues. Mol Cell, 64(5), 982-99

Printed in Great Britain
by Amazon

62465295R00131